Killer Apes, Naked Apes,
& Just Plain Nasty People

KILLER APES, NAKED APES, & JUST PLAIN NASTY PEOPLE

The Misuse and Abuse of Science in Political Discourse

Richard J. Perry

JOHNS HOPKINS UNIVERSITY PRESS

Baltimore

© 2015 Johns Hopkins University Press
All rights reserved. Published 2015
Printed in the United States of America on acid-free paper
9 8 7 6 5 4 3 2 1

Johns Hopkins University Press
2715 North Charles Street
Baltimore, Maryland 21218-4363
www.press.jhu.edu

Library of Congress Cataloging-in-Publication Data

Perry, Richard John, 1942–
    Killer apes, naked apes, and just plain nasty people : the misuse and abuse of
science in politics / Richard J. Perry.
        pages cm
    Includes bibliographical references and index.
    ISBN 978-1-4214-1751-6 (hardcover : alk. paper) — ISBN 978-1-4214-1752-3
(electronic) — ISBN 1-4214-1751-0 (hardcover : alk. paper) — ISBN 1-4214-1752-9
(electronic)   1. Sociobiology.   2. Anthropology—Philosophy.   3. Political
anthropology.   4. Science—Philosophy.   5. Science and civilization.   I. Title.
    GN365.9.P47 2015
    304.5—dc23   2014043205

A catalog record for this book is available from the British Library.

*Special discounts are available for bulk purchases of this book. For more information,
please contact Special Sales at 410-516-6936 or specialsales@press.jhu.edu.*

Johns Hopkins University Press uses environmentally friendly book materials,
including recycled text paper that is composed of at least 30 percent post-
consumer waste, whenever possible.

For Ali.

In love with you forever.

Dedicated to

Alice Pomponio

August 4, 1952–June 19, 2012

Loving wife, loving mother.

Internationally honored Professor of Anthropology

Contents

Preface

I'VE BEEN A STUDENT OF ANTHROPOLOGY SINCE THE LATE 1950s when in high school, I first discovered what it was. I studied it in college and in graduate school, taught it as a college professor for more than thirty years, and was married to a wonderful anthropologist for two decades. If any life has a theme, I guess mine has been anthropology. One could do worse.

I began this project when our son, Gregory, a college student, reacted with outrage to an assigned reading that seemed to promote biological determinism. When he shared his written response with me, my own reaction was similar to his. Having retired a few years earlier after teaching anthropology for more than three decades, I'd pursued various other interests since that time. My first thought was, "I can't believe they're still pushing this stuff."

I should have known better, of course. I had published a critique of sociobiology back in 1980, and a few years ago, I'd written a book on "race" and racism that dealt at some length with the history of biological determinism, eugenics, the misuse of IQ tests, and other abuses of biology in the pursuit of social and political agendas. In retirement, though, I'd paid less attention to the issue. (Following such scholars as C. Loring Brace, the word "race" here will always appear in quotes, reflecting the general view that the concept lacks scientific validity.)

In retrospect, I recall various articles in the popular media about such things as evolutionary preferences for dating partners and references to "alpha males" and even noted some biological determinist du jour showing up once in a while in the *New York Times Sunday Magazine*. But somehow I'd brushed over such things and moved on to items that seemed more engaging—the Sunday crossword puzzle or eight ways to cook chicken thighs. Now I felt like a reverse Rip van Winkle who wakes up after a long slumber to find that nothing has changed.

Despite my optimistic assumptions, it was evident that biological determinism—in whatever guise or terminological trappings—isn't dead after all. It's not even appreciably weaker. I had naively assumed that since its regenerative tendrils had been trimmed and discredited numerous times over the past decades, it had finally wilted as a serious enterprise. Wrong. In fact, it has continually regenerated, and it's still very much with us. It's remained a constant, if noxious, part of the intellectual ecosystem, an unwelcome assertive guest who stays way too long at the party.

If we hope to take some comfort in the thought that these old ideas, having been discredited so many times, no longer have much influence, we only need to look at the news of the day. One obvious question is why. The object of this exercise is to answer that question.

We'll address these and many related issues in the following pages. For the rest of this book, this is what we're in for.

I HAVE MANY PEOPLE TO THANK for their help and support though the process of writing this. Foremost among them is my son Gregory who, despite the demands of his own studies as a college student, patiently read through several drafts, suggesting corrections, additions, and more than a few deletions. Whatever value this book may have, his tireless efforts and encouragement have enhanced it greatly.

I want to thank Paul Lawrence Farber, Distinguished Professor of History of Science, Emeritus, at Oregon State University, who offered many helpful suggestions. And thanks to Robert J. Brugger, senior editor at Johns Hopkins University Press, who patiently and skillfully navigated the project through perilous waters.

I'm especially indebted to friends and colleagues who listened patiently to ideas, offered some of their own, and suggested readings—generally at our regular Thursday night get-togethers at the local Best Western. Special thanks to Shinu Abraham, Mindy Pitre, Adam Harr, and Wendi Haugh of the Anthropology Department at St. Lawrence University; Karen Dillon O'Neill, Leah Rohlfsen, Karen Gagne, and

Dan McClaine of sociology; Juan Ponce-Vasquez, Elun Gabriel, and Donna Alvah of history; Sandhya Ganapathy of Global Studies; and Karen Johnson-Weiner of anthropology at SUNY Potsdam. And to Jenny Hansen of philosophy at St. Lawrence, who actually read an earlier version of the manuscript. Although they've indulged and supported me, they're not responsible for any errors or shortcomings in the following pages.

Finally, to my wife and partner, Alice Pomponio, a beloved colleague of many of these folks, who passed away in June 2012. This is for her.

Killer Apes, Naked Apes, & Just Plain Nasty People

WHERE ARE WE GOING WITH THIS?

W E'RE ABOUT TO TAKE A BRIEF LOOK AT THE HISTORY OF biological determinism in its various forms over the past century and a half. To start, we'll consider some of the ideas Europeans of an earlier time held about human traits and differences and what they believed caused or cultivated these characteristics. We'll then follow the appearance of biological nuances and themes in these ideas beginning with the European Enlightenment of the eighteenth century, when biology began to develop as a science.

The tale continues through the rise of the eugenics movement and the upsurge of overt racism in the nineteenth and early twentieth centuries, when the focus shifts more directly to the United States and, to a lesser degree, Great Britain and Germany. We'll examine later arguments and assertions about "human nature" that drew inspiration from the advancing studies of human evolution, ethology, and genetics, eventually leading to the rise of sociobiology in the 1970s. The process continues to the present, culminating with a focus on the current enterprise of evolutionary psychology and its social and political implications.

In the various political and philosophical contexts of each historic phase, biological explanation has often resonated to justify deeply conservative points of view. The idea that things are the way they are because of inherent biological factors, and therefore are not susceptible to rapid change, has an inherently conservative appeal.

Could it be that the apparent upsurge in media interest over the past decade or so is due to advances in science that have lent more support to these assertions? The task here is to explore that possibility. This is not an exercise in seeking villains or impugning motives. The focus is on the ideas and assertions themselves. Where do they come from? How do they stand up against scientific research?

It will probably surprise no one that my own conclusions are that they don't—conclusions I share with many who have explored the question over the years. But I accept that it's necessary to make the case and not simply assert it. This book presents the debate both from a historical perspective and in based on we know so far.

The point here is to examine the ideas themselves rather than the people who've held them. It's the assertions, their scientific validity, and their social implications and consequences that we're after. In the end, the reader will have to judge the merits of the arguments.

As we'll see in subsequent chapters, arguments claiming biological determinism on the basis of "new" scientific findings are not new, nor are they based on scientific findings. They've reappeared in every era from the early days of "scientific racism" to the present—sometimes controversial, at times widely accepted. This is why a historical perspective is so important. In every era, proponents of such views have alluded to the scientific research of the day, but actual research has failed to support them. In most respects, these arguments have amounted to ideological assertions rather than scientific discoveries—a matter of "faith without proof." It often seems to be a short step from assertions of "what's evolved to be" to "what's meant to be." We'll discuss this in detail in the following chapters.

It might seem odd or even inappropriate to link alleged "scientific research" with social ideology. It's been a strange romance indeed. Yet, in some respects, these different domains can't seem to stay away from one another. The ideology side of the relationship, though, has been the ardent pursuer, while science, concerned with other things, hardly seems to notice. Ideally, as we know, a fundamental requirement of science is objectivity. But as we've noted, the trappings of science can lend credibility to a philosophical perspective.

These multiple versions of biological determinism over the years have led to serious harm to many people. They promote a worldview that emphasizes competition, a Hobbesian war of all against all, the idea that somehow, assisting the weak and disenfranchised is detrimental to progress.

At heart, it's the same old social Darwinism, now often expressed as neoliberal economics; although there is nothing really "neo" about it, and the only liberal aspect is liberation from any sense of collective social responsibility. It would be discouraging enough if such conclusions had scientific validity. If this is not the case, it makes them far worse, since they're essentially unconstrained by either the lack of sound data or the limits of public credulity, which, as we know, can be very flexible.

This issue is of crucial importance. Whatever motivations the proponents of these ideas might have, once these assertions enter the public discourse, they become tools for those who oppose change and reform or for those who wish to dismantle social programs they consider to be "going too far" in helping the disadvantaged, being soft on immigrants, or allowing the poor to reproduce. It would be justified to consider many of these initiatives as going beyond conservatism to the point of being reactionary. Rather than guarding the status quo, they would dismantle aspects of it to return to an earlier era—whether real or imagined.

Scholars and researchers have engaged in these arguments for decades but usually within academic venues. Specious claims of a gene for this or that are likely to enjoy space in the *New York Times* science pages, but the sound research nestles deep in the quiet corridors of the science library. As Mal Ahern and Moira Weigel put it, "within the academy, many of EP's methods have been questioned and even discredited. Yet even as academics continue to point out its flaws, the mainstream media have increasingly accepted evolutionary psychology as a mode of explaining human behavior."[1]

This book is for those who don't really have the time or inclination to wade through scholarly journals—for people whose daily lives are filled with other demands, but who often hear about genes for this or that kind of behavior, or tendency, or personality trait and wonder about the issue. On the other hand, the footnotes offer ample documentation and references for all major points. Anyone who'd like to pursue any topic more deeply should find it easy to do so.

Nowadays, as in the past, biological determinism claims the status of science. But science, of course, is more than a magic word that one invokes to ward off critique. Science involves evidence; what we call data. Science involves constructing hypotheses that one can test against such data. The most recent versions of biological determinism, like their counterparts in the past, fail to meet these criteria.

It doesn't really matter whether the people promoting such policies truly accept pseudoscientific assertions at a personal level. The important issue is that the assertions become widely available for general application. The only remedy is to present critiques and counterarguments in the same public arena.

To be clear, this is by no means a critique of biology or of any other legitimate science. On the contrary, it's a critique of the misuse of biology,[2] the co-opting of scientific camouflage to lend an aura of authority to a social ideology.

The point of the exercise is to evaluate the claims of scientific validity that proponents of biological determinism have made at each historical juncture. Although evidence has not supported such claims, it's important to examine the ideas, consider the propositions and assertions on which they've rested, and assess the state of science in the times in which they've appeared. In each case, as we'll see, such claims have been unfounded even in their era's scientific contexts.

What we're dealing with, essentially, has not been science, but ideology presented as science. And although the various branches of genuine science have advanced tremendously over the past century and a half, arguments for biological determinism have remained pretty much the same.

At this point, we should discuss what we mean by *biological determinism*. In a literal sense, biological *determinism* would be an absurdity, if we take the extreme interpretation that biology alone determines all of an organism's actions. The closest example in nature might be social insects—ants and bees. But external factors lead even these well-organized creatures to alter and adjust their behavior. There seem to be few, if any, examples of some sort of bio-robots whose behavior responds rigidly and exclusively to predetermined inner drives. No one, to my knowledge, has ever seriously attempted to apply such a model to the human experience.

Many, however, have argued that our biology, our genetic makeup, strongly influences, molds, guides, impels, and limits how we feel, how we react, and what we're likely to do. At that level, alleged biological influence is closer to being "determinative"—that is, "having the power or tendency to determine [the outcome of something]."[3] Few writers, I suspect, would dare to inflict the term *biological determinativism* on innocent readers. The original ten-syllable term is awkward enough. Let's stipulate, then, that the use of *biological determinism* here is intended to convey the meaning as *determinative* rather than *absolutely deterministic*.

THERE'S A HISTORY HERE

Regarding the appeal of biological determinism to conservative thinkers, Gunnar Myrdal put it succinctly back in 1944. "The individual and society can . . . according to the liberal, be improved through education and social reform. The conservative, on the other hand, thinks that it is 'human nature' and not the environment which, on the whole, makes individuals what they are."[4] He adds in a footnote on the same page that "it is hardly possible to be a true biological determinist and yet a political liberal."

This is why the issue is far more important than a series of squabbles among academics. These debates over the decades have justified a variety of political policies—policies that have had profound material effects on real people, on the ground, in real time.

Given the complexity of the human mind, we can always find some exceptions to the more general linkage of biological determinism to political orientation. As we know, humans are quite capable of holding inconsistent, and perhaps even contradictory, views at the same time.

The noted linguist Noam Chomsky, for example, has argued that innate grammatical structures common to all languages are inherited biologically.[5] Chomsky has even argued that most ideas preexist in the structure of the brain rather than occurring purely as a result of learning.[6] He has consistently emphasized the importance of biologically determined mental structures over learned experience. Yet Chomsky has also been known over the years for espousing a variety of liberal causes, opposing racism, sexism, the Vietnam War, and other military policies. He has tended to shun the term *liberal*, however. According to one source, he prefers the term *libertarian socialism*,[7] whatever that might mean.

Chomsky offers one example of the myriad and sometimes puzzling perspectives that have often appeared in the political arena. We might also consider Richard Dawkins, a staunch biological determinist.[8] Beyond his prominence as an outspoken atheist, Dawkins took stands against the United States war on Iraq and many of President George W. Bush's policies in general. He has described himself as a long-standing voter for the British Liberal Democrats. Apparently there's no particular gene for consistency.

The eminent French anthropologist Claude Lévi-Strauss, like Chomsky, discussed underlying structures in human social and cultural life.[9] Unlike Chomsky, however, Lévi-Strauss placed great importance on learning and experience. He tended to see structures as means of managing, reconciling, or even disguising many of the difficult or contradictory elements of lived human experience. In this regard, despite the coincidental use of the term *structure*, Lévi-Strauss was far from being a biological determinist in the sense of either Chomsky or Dawkins. He was also active in socialist politics, especially early in his career.[10]

As we can see, those who have engaged in the debate have not always fallen neatly into standard categories. Early in the twentieth century, many public figures who were considered liberal in their time— Margaret Sanger, George Bernard Shaw, or Theodore Roosevelt,[11] for example—were enthusiastic eugenicists. Few liberals today would embrace such a position. One reason, of course, is that between the 1920s and the present era, the world witnessed the nightmare of the Holocaust—an atrocity that grew directly out of eugenic thinking.

When we consider the issues of human behavioral flexibility versus biologically inherent patterns, and then associate these with people who favor change and reform versus those who favor stability and the status quo, opinions are not likely to show anything like a random distribution. The marbles all pretty much roll to one end or another. Although variations and outliers exist, liberal biological determinists these days are probably about as rare as socialist Republicans.

In the contemporary United States, the recent neoliberal economic and political emphasis on free markets, deregulation, and defunding social assistance programs has led to drastic funding reductions for the Head Start preschool program designed to assist disadvantaged children; refusal to fund long-term unemployment payments for those continuing to look for work; the cutting of food stamp assistance for those unable to earn enough to feed their families and the Meals on Wheels program which delivers food to elderly people unable to leave their homes; and continued assaults on other government programs intended to help people in need. Although the federal government, through the Affordable Care Act, made medical care available to people in financial need through Medicaid, governors of more than half the states, including many of the poorest, refused to make federally subsidized Medicaid available to their citizens. It would be difficult to assess how many people have died as a result.

These developments are reminiscent of old-fashioned social Darwinism with its theme of "survival of the fittest." Assertions of a biologically scripted human nature have been a consistent theme. In various historical eras, these arguments have risen to justify colonialism, slavery, gender discrimination, neoliberal economic policies, and in general, the status quo of the day.

The central themes of sociobiology and evolutionary psychology, stressing the competition of genes for survival and the fixity of genes for behavior allegedly "selected for" in the Pleistocene, perpetuate elements of this old idea with new vocabulary. A century and a half ago we heard that traits or characteristics of particular ethnic groups or social classes were inherent—"in the blood." Hence, attributes such as criminality, promiscuity, or a propensity for poverty were not subject to change through social assistance programs. Nowadays, we hear of genes at the individual level that shape our destinies. We've had reports of genes for personality traits, for dating preferences, even for political tendencies.

We might assume that such ideas, refuted so many times in the past, would have gone out of fashion by now. We'd be mistaken. In 2013, the old issue of inherited "racial" differences in IQ, so familiar to those of us who can remember the 1960s, reared up again in a report written for the conservative Heritage Foundation addressing immigration laws. The report asserted that the IQ scores of Latinos are significantly lower than those of "whites" and are certain to remain so for several generations, because of genetic limitations. In 2014, an opinion piece in the *New York Times* made a case for an inherited biological propensity to accumulate wealth. The evidence? Continued prosperity in the same family lines over generations.[12]

In 2014, the prolific journalist Nicholas Wade published *A Troublesome Inheritance*, arguing that alleged behavioral differences among "races" have a genetic basis and that major "races" differ in their inherent capacities.[13] Some readers might feel their jaws drop slightly to read assertions this far into the twenty-first century that some "races" are less capable of civilization than others. In a review of the book, the distinguished biologist H. Allen Orr noted that "hard evidence for Wade's thesis is nearly nonexistent."[14] We'll discuss the fallacy of the concept of "race" in detail in chapter 3.

WHAT'S IN A WORD?

Many terms that play an important part in this and later discussions have political implications. It's useful to consider not only what these terms mean but also what they have meant, especially because from one era to another, many of them have taken on different, sometimes contradictory meanings.

We've already mentioned *conservatism*. But even this apparently straightforward term subsumes a range of meanings. Politicians may describe themselves as "fiscally conservative" but "socially liberal"— apparently meaning that they don't much care what people do in their private lives, as long as the government doesn't have to pay for it.

And clearly, not all conservatives are biological determinists. Some who would define themselves as conservatives deny evolution altogether and espouse a biblical version of creation. And if that weren't confusing enough, many conservatives oppose environmental conservation, seeing ecological activists as "tree huggers" carrying out a "war on coal."

We also hear about "ultraconservatives," people whose opponents might even refer to as "reactionary." (It's rare to hear of anyone who actually claims to *be* a reactionary.) That label often refers to those who advocate change, not toward something new but, rather, to the way things once were—or, at least, the way they should have been, even if they weren't.

Nowadays, many people on the political left of the spectrum identify themselves as *progressive* rather than as *liberal*. Indeed, the mission statement of the *Progressive*, a major national publication, expresses a desire "to change the world, for the better. . . . Our reporting and analysis take on the modern-day robber barons, and champion peace, civil liberties, equality, and justice."[15]

A century or more ago, however, the term *progressive* had a rather different slant. Social Darwinists of the early twentieth century embraced the principle that the mechanism of progress was the survival of the fittest. Social Darwinist Herbert Spencer in the late nineteenth century argued that progress in nature comes about through competition, with the strongest (the most "fit") surviving. In such a view, charities and social programs to help the needy only interfere with the natural process of weeding out the weak. In that era, robber barons (who would probably have preferred the term *captains of industry*) epitomized progress in the minds of many. Equality might have entailed the right of all to compete, but not for all to share equal social benefits.

The term *liberal* has also undergone a shift in meaning. One thing that made the robber barons possible was nineteenth-century liberal economics—that is, business liberated from government regulations or constraints. It was good old-fashioned laissez-faire capitalism, the free market. As we'll discuss in later chapters, the current term *neoliberal* harkens wistfully back to those days.

We don't hear much anymore about the other liberalism of the twentieth century. That would have been the liberalism of Theodore

Roosevelt, who broke up industrial trusts and monopolies; Franklin D. Roosevelt, who pushed through public works and Social Security; Harry S. Truman, who racially desegregated the military; or Lyndon B. Johnson, who pushed through civil rights legislation. By the 1970s, opponents of such programs—which conservatives often referred to as the "liberal agenda"—had managed to create a public image of liberals as busybodies trying to interfere with people's personal freedoms. This became more feasible once a generation or so had gotten used to the benefits of liberal programs and had begun to take them for granted. As a consequence, many who would once have been "proud liberals" tended to tiptoe away from the term by the late 1970s. Ironically, for many of those farther to the left, the term also came to mean too much willingness to compromise—an exasperatingly soft commitment to important social causes.

We also have a range of other, perhaps lesser terms—*leftist, radical,* and so on—that subsume a range of possible meanings. And there's probably no value in spending time on the grossly overused epithets *fascist* and *Nazi*. Here, *conservative, liberal,* or *progressive,* will mean the contemporary, commonplace understanding of the terms, rather than what they meant in the past.

THIS BOOK MAKES THREE MAJOR POINTS. The first is that human behavior, while limited in certain respects by our biological heritage, is not fixed, programmed, or biologically determined. Human history and cross-cultural studies amply demonstrate the human capacity for change.

Second, the arguments that continually appear claiming biological determinism on the basis of "new" scientific findings are not particularly new, nor are they based on scientific findings. They've reappeared in every era from the early days of "scientific racism" to the present. This is why the historic approach is so important. In every era, biological determinists have alluded to the scientific research of the day, but actual research has not supported them. In most respects, these arguments have been ideological assertions.

Third, the issue is of crucial importance to society because, whatever the proponents' motivation, once these assertions enter the public discourse they provide tools for those who oppose change, or who wish to dismantle social programs they feel to be "going too far" in helping the disadvantaged. It doesn't matter whether proponents of these ideas accept them at a personal level. The more important issue is that the assertions become available for use. I would never advocate suppressing such assertions. The crucial point is that they need to face counter-arguments within the same public arena. The appropriate response is refutation.

We can take a closer look at the historical processes that have brought us to where we are now. Complex dynamics have fed into the flow of events and ideas. As scientific knowledge has increased, social, political, and economic forces have had their effects, with ideologies and vested interests often deflecting the trajectory of public opinion and political policies. Many of us continue to have confidence that increasing knowledge and advances in science are bound to result in progress. We are, after all, still children of the Enlightenment—at least most of us. But in some aspects of life—relationships among categories of people, views of what society should be, social and political strategies to promote advantage—evidence of progress is sometimes far from clear.

Chapter 1

Don't Get Me Started

FOR AT LEAST A CENTURY AND A HALF, A PERSISTENT AND thriving academic enterprise has drawn energy from the urge to account for the complexity of the human experience in biological terms. As far back as we can tell, of course, humans have always sought various ways to account for themselves and, even more entertainingly, to account for the strange ways of others. These attempts have invoked a range of causal factors ranging from climate, diet, and geography to the imponderability of the Creator. What sets this particular endeavor apart, though, is the biological theme.

For this to have happened in the first place, of course, one needed biology—that is, an intellectual, nontheological framework based on scientific observation, however halting or inaccurate that framework may have been during its early days. The science of biology has a historical beginning. Like other scientific endeavors, biology as we know it has its roots in the European Enlightenment of the eighteenth century, although it did, of course, draw on knowledge that was far more ancient.

The Enlightenment, most scholars would agree, marked a departure from a reliance on theological explanation to account for human affairs and other events large and small in the world. God's will could cover a lot, but in many quarters, intellectuals felt the desire for more specific, immediate, and fine-tuned analysis.

Some Enlightenment scholars hedged a bit as to whether God exists. But those who did assume the presence of a divine being tended to see His influence in having set things in motion in the first place, presumably establishing laws and principles by which the universe operates. Once that had been done, many assumed, the most important questions rested on what these laws might be and how they function. From that perspective, the best way to address those questions was to observe the workings of the world directly and objectively, to describe them meticulously, to measure their manifestations accurately, and to strive toward some logical, testable hypotheses capable of accounting for them—and, perhaps, even to acquire enough knowledge to predict their results in future cases. For these questions, ancient scriptures were of little help.

As anyone in the United States who's been paying attention lately can attest, the principles of the Enlightenment have not eclipsed medieval thinking in all quarters. We still have politicians (duly representing their constituencies, one can only assume), proclaiming that catastrophic weather events are expressions of God's wrath because of sinful human behavior. In 2013, Congressman Joe Barton of Texas assured his audience that the Great Flood of biblical fame proves that human activity doesn't cause climate change. And U.S. Rep. Paul Broun of Georgia has proclaimed that the idea of evolution comes from the "pits of Hell." He also included the science of embryology in his infernal condemnations. Who knew? Congressman Broun holds a degree in medicine. He was also a member of the House Committee on Science, Space, and Technology. Louisiana governor Bobby Jindal, who also has expressed misgivings about this evolution stuff, was a Rhodes Scholar and an advisor to the federal government on health policy before entering the statehouse in Baton Rouge. Recent surveys indicate that compared to several decades ago more of the U.S. population now believe that the earth is only a few thousand years old and that dinosaurs once coexisted with humans.

Despite such anachronisms, however, secular scientific thinking since the Enlightenment has held sway, to the extent that we now have far more hybrid cars than ox carts, and more people today rely on cell phones for communication than on messengers on horseback galloping over muddy roads.

This does not, of course, mean that science can give us the answers to everything. We still face all sorts of ambiguities, moral dilemmas, and emotional issues. Probably we can all agree that much of human behavior—what we do—arises from what we know. And what we know generally means what we've learned and how we interpret that information, which in turn affects how we choose to act (or not act) on our knowledge.

Of course, the capacity to perceive, to process, and to interpret experience has a biological basis. A dauntingly complex series of receptors, neurons, pathways, synapses, and so on allow humans to perform these operations, which they do millions of times a day unless they suffer some impairment of their standard equipment. More and more, thanks to scientific research, it's become possible to identify many such disabilities and to locate their causes—whether genetic, chemical, or a result of physical injury.

This standard biological human equipment is a product of evolution over millions of years. It does not, however, determine in any specific way just *what* an individual may have learned in the course of a lifetime, the myriad experiences a person might have—whether traumatic, life enhancing, or innocuous—or how she or he might interpret or even be subconsciously affected by them. These innumerable factors, clearly, are major influences on human behavior. This is one reason why humans display cultural diversity.

By culture, we mean learned, patterned behavior—behavior that people have come to share as a result of living among one another—as opposed to any innate or biologically fixed characteristics. This is not to say that people have no such biological characteristics; only that these characteristics are distinct from culture.

Many proponents of a strong biological explanatory framework to account for human behavior emphasize the biological (or genetic) aspects of human life over the messier, more complicated, unpredictable, and often chaotic experiential aspects of human existence. There's one major problem with this. We hardly need to point out that human lives are extremely variable. But is the same true of human biological makeup, aside from pathologies? Not so much. All of the evidence to date underscores the essential genetic unity of the human species. (We'll discuss this further in chapter 3.) Many adherents of biological explanation apparently would disagree, however. Many of them have argued, and have attempted to demonstrate, that observed behavior patterns among human beings are predominantly expressions of genetic differences.

In some cases, these arguments have focused on differences among individuals. In other cases, they've addressed alleged differences among populations or other demographic categories. Some of these approaches have been rather cautious, even tentative. Others have been more emphatic, perhaps even extreme, to the extent that we could fairly characterize them as assertions of biological determinism.

Many scholars have objected to the word *determinism*, while nonetheless asserting strong biological influences on behavior, what we might call a genetic directedness, or perhaps a biological scripting, of behavior. As we'll see in later chapters, the semantics have been a little slippery. Within this spectrum, various versions of racism, as well as classism and sexism, have found a comfortable home.

This is not to say that all forms of biological explanation are racist, classist, or sexist. Yet to the extent that such approaches attribute innate and immutable biological differences affecting behavioral attributes and abilities of populations (whether one defines these populations in terms of ethnicity, geography, social class, or sex), it's not difficult to see an epistemological compatibility.

HUMANS AND "NATURE"

Assuming that we can all agree on what we mean by "human," it might, nonetheless, be worth considering what we mean by "nature." Most inclusively, at least in European and American thought, nature has included not only living but nonliving aspects of the earth such as volcanoes, steppes, mountain ranges, oceans, and weather patterns. The term *nature* has carried a sense of powerful forces. Some even visualize nature as a stern, unforgiving mother whom one can't fool, and with whom one should not mess.

For present purposes, though, our focus will be on the biomass, that component of nature that consists of living things, and particularly on the realm of animals. Within that realm our primary concern is with what we might arrogantly refer to as the higher animals: those species that are more like us than, for example, "bugs" of various sorts. (Although as we'll see in a later chapter, ants and bees also have had a significant impact on some views of human nature.)

There are a number of possible ways in which we could visualize the relationship of humans to nature (or to the *rest* of nature, as some would see it). As many scholars have noted, Judeo-Christian traditions have set humans *in opposition* to nature. In the book of Genesis, God famously gave Adam and Eve dominion over the fowl of the air, the beasts of the earth, the fish in the sea, and pretty much everything else (except one particular fruit tree—but that's another story for another time). The implication, clearly, was that Adam and Eve represented a very special type of being indeed.

Yet as we know other perspectives have pervaded in various parts of the world. People in some societies have believed that humans once were animals among other animals and underwent transformative adventures to arrive at our present state—not always necessarily viewed as superior, by the way. Others, conversely, have believed that animals once spoke and lived like humans but, again, underwent experiences that caused them to change. Still others have believed that we humans are creatures of nature equivalent to other species. In that vein, in some systems of belief the idea of nature as an entity distinct from humanity makes no sense. At least one major world religion has souls traversing numerous life-forms through a series of reincarnations. Concerned environmentalists in our own society urge that we adopt a sense of shared existence with other species and accept the implicit responsibility for our mutual well-being.

To risk oversimplification, we could summarize the possibilities as a view that humans are separate from nature and dominant over it; a view that humans are *not* separate from nature but a part of it; or a view that humans are separate from nature but susceptible to it, reliant on it, and responsible to help maintain it (or at least, to refrain from destroying it). At the level of sterile logic, these may appear to be mutually exclusive, but in fact they blend nicely in the simmering stew of human discourse. We can see aspects of all of them among many of the arguments for explanatory frameworks that emphasize biological causal factors in human life.

Beyond the concepts of "humans" and "nature," we also have the idea of "human nature": that is, the inherent qualities of *Homo sapiens*, whatever they might be. This might seem to be a matter of quibbling over minor details, but in fact, it has been a rather contentious issue. On one side, some would argue that beyond the basic evolved, anatomical biology of the species itself, a fundamental aspect of human nature is the ability—in fact, the need—to learn, thereby lending an inherent flexibility to behavior patterns. At the other corner of the ring, some would emphasize inherent, fixed aspects of human nature, in the sense that human behavior follows strong biological imperatives. Such biological compulsions allegedly are the results of genetic predispositions that were selected for during the course of evolution. We'll revisit this issue more fully in later chapters.

IT WASN'T ALWAYS ABOUT BIOLOGY

In European thought, some of the most concerted endeavors to explain human differences in biological terms and to try to do something about them arose in the nineteenth century, with the work of Charles Darwin's cousin, the statistician and pioneering eugenicist Sir Francis Galton. Sir Francis will receive more of the attention he deserves in the next chapter. Before that, however, references to biological or physical traits in most writings generally amounted to little more than fleeting mentions or passing observations.

Living in this century, many of us might assume that the idea of innate biological human differences is self-evident. For most of recorded history, though, writers who addressed the human condition saw things differently. In considering the known (and sometimes imagined) variety of humanity throughout the known world, many early writers began with an assumption of a common stratum of fundamental shared human nature. Starting from this basis, the most logical causes for different human characteristics would be environments or historical experiences.

Herodotus, a Greek historian and early ethnographer of sorts who lived around the fifth century BCE, traveled over much of the known world of his time. He described peoples whose customs and traits would have seemed bizarre to his audience at home. He speculated about causes and even suggested that some of these populations were superior to his own in some ways. He admired Egyptians in particular. Although Herodotus noted and reported physical differences, he generally attributed these to environmental factors such as the climate or the quality of the air. The anthropologist Clyde Kluckhohn wrote that in Herodotus's view, "men were not classified as black or white but as free or servile."[1]

Herodotus was fairly typical of his time in seeing human differences as the results of experience rather than as biologically innate, although some writers seemed to feel that once acquired, traits could pass down to future generations. Herodotus quotes the views of King Cyrus of Persia on the power of environment to affect human character: "soft lands tend to breed soft men. It is impossible, he said, for one and the same country to produce remarkable crops and good fighting men . . . and they chose to live in a harsh land and rule it rather than to cultivate fertile plains and be others' slaves."[2]

The Greek physician Hippocrates, writing in the same era, expressed a similar perspective. In describing the inhabitants of the Phasis River area, he wrote that

> the country is marshy and warm and well watered and thickly clothed with vegetation, and there is heavy and violent rainfall there at all seasons. . . . The waters they drink are warm and stagnant and putrified by the sun. . . . And mist envelopes the country as a result of the water. For just these reasons the Phasians have their bodily form different from those of all other men. For in stature they are tall, in breadth they are excessively broad, and no joint or vein is seen upon them. Their complexion is yellow as if they had the jaundice. Their voice is the deepest of all men's because their atmosphere is not clear but foggy and moist. And for bodily exertion they are naturally disinclined.[3]

For early Christian writers, a common human nature was the result of divine creation, God's will. Augustine of Hippo, writing a millennium or so after Herodotus and Hippocrates in the fifth century CE, was quite explicit: "Whoever is anywhere born a man, that is, a rational, mortal animal, no matter what unusual appearance he presents in color, movement, sound, nor how peculiar he is in some power, part, or quality of his nature, no Christian can doubt that he springs from that one proto-plast." Further, "if they are human, they are descended from Adam.[4]

Thomas Aquinas in the thirteenth century wrote that "all men born of Adam may be considered as one man, inasmuch as they have one common nature, which they received from their first parents."[5] Implicit in this is the sense that this nature, while divine in origin, is also inherited.

The Arab scholar Ibn Khaldun, who was born in Spain in 1332 and spent much of his life there, also traveled throughout the Middle East. His observations led him to place great importance on environmental factors in molding human characteristics. Heat, he believed, stimulated the "animal spirit." He observed that "when those who enjoy a hot bath inhale the air of the bath, so that the heat of the air enters their spirits and makes them hot, they are found to experience joy. It often happens that they start singing, as singing has its origins in gladness."[6] He also noted that "the desert people who lack grain and seasonings are found to be healthier in body and better in character than the hill people who have plenty of everything. Their complexions are clearer, their bodies cleaner, their figures more perfect and better, their characters less in-temperate, and their minds keener as far as knowledge and perception are concerned."[7]

There's little evidence that Ibn Khaldun's ideas ever directly affected European thought at the time, since none of his works were translated into a European language until the nineteenth century. As we'll see, though, many European thinkers well into the early years of that century also continued to place great importance on environmental, noninnate factors in shaping and modifying the basic, common human material.

The invention of the printing press with movable type in 1445 en-hanced the sharing of ideas among a wider audience and stimulated scientific discourse in general. Not everything that appeared in print, of course, was up to the best standards of science.

Some of the earliest printed descriptions of exotic varieties of human-ity turned up well before that in Europe in the fourteenth century, in a book of travelers' tales published under the name Sir John Mandeville. The book was supposedly an account of travels through the Middle East, Asia Minor, the general Mediterranean region, and the remote reaches of Asia. The tales offered a rich collection of ethnographic accounts of some of the peoples who actually inhabited those parts of the world, but they also included fanciful accounts of creatures that would have been bizarre by any standard. Much of the jaw-dropping material that "Sir John" presented was drawn from European folktales that had enthralled wide-eyed children around kitchen hearths for generations. The work entertained a wide reading audience for centuries.

Most of the accounts of actual populations tended to focus on different customs, whether accurately or not, rather than on different physical characteristics. At the same time, this engrossing volume also offered tales of human-like creatures who ran through the trees like squirrels; others whose mouths were tiny holes in their faces; and still others with one huge foot that came in handy, one might say, to hold up on a hot sunny day as a sort of anatomical parasol. The book describes "folk of gret stature as geaunts, and thei be hidouse to loke upon. And thei had but on eye, and that is in the myddle of the front."[8]

By the sixteenth century, Europeans had developed a more realistic view of other populations, having experienced more intimate and prolonged contact with peoples in distant parts of the world—though not always on a friendly basis. Many of these contacts came about in the search for resources, possible trade opportunities, and increasingly, as colonial ventures. Increased contact inspired still more speculation about the differences among people of various regions.

Oddly enough to modern readers who are all too familiar with "racial" explanations, as we've noted, such factors as skin color or hair form were not often European writers' primary topics of interest. This is not to say that Europeans refrained from aesthetic judgments, of course. On this particular issue, historian Pietro Martire (aka Peter Martyr) in the early 1500s argued that differing ideas of beauty and repulsiveness teach us "how absurdly the human race is sunk in its own blindness, and how much we are all mistaken. The Ethiopian considers that black is a more beautiful color than white, while the white man thinks otherwise. The hairless man thinks he looks better than the hairy one. . . . It is clearly a reaction of the emotions and not a reasoned conclusion that leads the human race into such absurdities, and every district is swayed by its own taste."[9]

Some negative assessments of far-off people's alleged behavioral properties provoked dissent. In the 1500s, the French philosopher and social commentator Michele de Montaigne took his fellow Europeans to task for holding arrogant views of barbarians. Montaigne drew his examples primarily from native peoples of a place we have come to know as Brazil, a few of whom he had encountered in Europe.

Montaigne used his defense of these "barbarians" as a device for critiquing what he felt were the vices, cruelties, and other aspects of decadence of his own society. His discussion rested mostly on the relative values of customs, behaviors, and morality. Montaigne noted that "no man's brain has yet been found so disordered as to excuse treachery, disloyalty, tyranny, and cruelty, which are our common faults. We are justified therefore in calling these people barbarians by reference to the laws of reason, but not in comparison with ourselves, who surpass them in every kind of barbarity."[10] In the same vein, "the savages who roast and eat the bodies of their dead do not scandalize me as much as those who persecute and torture the living."[11]

As we might expect, still other explanations of human differences at that time rested on theology. Spaniards in the Americas encountered civilizations that in many ways were more complex than their own, with extensive empires, paved roads, regional markets, magnificent temples, and other monumental structures. But the new diseases the Spanish brought with them soon killed the majority of these indigenous peoples, and enslavement and other oppressive policies destroyed even more. The diseases, given the state of knowledge at the time, were almost certainly unintentional. The policies, not so much. Many of these destructive colonial practices raised questions of morality among the Spanish hierarchy.

A major issue that arose at the time was whether native peoples in the Americas were truly human in the same sense as the Spanish. In the sixteenth century, this was not a question of evolution but of divine creation. Were the indigenous peoples of the Americas descendants of Adam and Eve as the Spanish and other Europeans were? Or were they the offspring of a different creation? Lengthy debates among theologians of the School of Salamanca ultimately led to the conclusion that they were, indeed, children of God with souls—which meant, therefore, that they should not be murdered or enslaved. Unfortunately, this had little effect the actions of Spanish colonists in the Americas, thousands of miles across the Atlantic.

BUT MAYBE THEY REALLY ARE DIFFERENT!

Although environmental factors remained important in explanations of human differences, we begin to see more suggestions of inherited biological qualities by the sixteenth century. Italian artist and inventor Leonardo da Vinci observed that "the black races in Ethiopia are not the product of the sun; for if a black gets black with child in Scythia, the offspring is black; but if a black gets a white woman with child the offspring is grey. And this shows that the seed of the mother has power in the embryo equally with that of the father."[12]

Some of the best-known arguments linking behavioral attributes to environment appear in the work of French philosopher Montesquieu in the early eighteenth century. Here, as in earlier writings, we find a sense of profound differences among populations, but the alleged behavioral rather than the physical characteristics of these exotic peoples still predominate in his discussions. Montesquieu seems to have no doubts about the relative values of such differences. But to account for them, he doesn't invoke their biological heritage but the nature of the climates in which these people live.

Montesquieu writes that men in the "whole of warm countries are, like old men, timorous; the people in cold countries are, like young men, brave." On the other hand, in "cold climates they have little sensibility for pleasure; in temperate countries, they have more; in warm countries, their sensibility is exquisite." And apparently when it comes to sexiness, climate also makes a huge difference. "From this delicacy of organs peculiar to warm climates it follows that the soul is most sensibly moved by whatever relates to the union of the two sexes; here almost everything leads to this subject." Is that all they think about? "In northern climates," by contrast, "scarcely has the animal part of love a power of making itself felt."[13]

We can find some apparent contradictions in Montesquieu's writings, but the deeper and perhaps more important implication is that, in his view, once again, profound human differences are not innate. They are not fixed in the very substance of these populations in a way we might think of as biologically inherited traits. They result, instead, from environmental conditions. A further implication is that anyone else who might come to be immersed in those same conditions, whatever his or her biological origins, would acquire the same characteristics as these lazy, fierce, unfeeling, or sexually obsessed folks Montesquieu describes. In fact, Montesquieu states that children of Europeans who are born and raised in India take on many of the characteristics of Indians.[14]

Studies of human anatomy, along with other sciences, advanced rapidly after the fifteenth century when printed works became more accessible, but nonbiological factors continued to be important in explaining differences. Physician Andreas Vesalius of Belgium in the sixteenth century noted that "the Germans, indeed, have a very flattened occiput and a broad head, because the boys always lie on their backs in their cradles. . . . More oblong heads are reserved to the Belgians . . . because the mothers permit their little boys to sleep turned over in their beds, and as much as possible on their sides."[15]

In the same era, French philosopher Jean Bodin dismissed simplistic environmental explanations for physical differences, noting the diversity of populations in what seemed to be comparable regions and altitudes. But like Montesquieu, he did see important environmental effects on character traits.

> The people of the South are of a contrarie humour and disposition to them of the North . . . the one is joyful and pleasant, the other sad; the one is fearful and peaceable, the other is hardie and mutinous; the one is sociable, the other solitarie; the one is given to drinke, the other sober; the one rude and grosse witted, the other advised and ceremonious; the one is prodigal and greedie, the other is covetous and holds fast;

the one is a souldier, the other a philosopher; the one fit for arms and labour, the other for knowledge and rest.[16]

Explicit assertions of innate differences among human populations became more common in the eighteenth century. The Scottish historian and Principal of the University of Edinburgh William Robertson, writing about native North Americans in 1777, describes these populations in derogatory terms that sound profoundly bigoted today. Regarding the inhabitants' stubborn resistance to forced labor, Robertson attributes this exasperating characteristic to an intrinsic "native indolence." He goes on to note that "this feebleness of constitution was universal among the inhabitants of those regions in America which we are surveying, and may be considered as characteristic of the species there."[17]

These were still pre-evolution times. Common views about the human past did not rest on the assumption that there had been a long progression from a lower, more primitive state. On the contrary, they assumed degeneration from a higher one. Thus, we find little evidence of the idea that native peoples represented earlier stages of development. More commonly, writers such as Robertson assumed that these populations had declined to their present miserable existence from a more glorious past.

Whether one embraced the idea of a biblical fall from the Garden of Eden or, in a more secular vein, one extolled the glories of classical Greek, Roman, and especially Egyptian civilizations or other ancient centers of high culture, a common assumption was that things had pretty much gone downhill over the ensuing centuries.[18] It seemed apparent to many intellectuals such as Robertson that the inferiority they perceived in "savage" populations was a result of their having degenerated. In the case of the Americas, this most likely was a consequence of distance in both time and space from earlier centers of the highest human achievement.

In the mid-eighteenth century, Swedish naturalist Carolus Linnaeus presented an early attempt to produce a scientific nomenclature of human types, offering a taxonomy that included descriptions of physical characteristics.[19] He described human types that included the four conventional continental "races": white Europeans, red Native Americans, black Africans, and brown (later yellow) Asians. But he also included troglodytes ("cave men") and satyrs. Despite such minor aberrations, the thrust of his work, and its most controversial aspect at the time, was to place humans within a natural continuum of plants and animals.

In a general sense, this reflected an older concept known as the "great chain of being,"[20] which most scholars of the time accepted as being more or less consistent with the idea of divine creation. It amounted to a means of sorting out the many creatures God had placed hither and yon throughout the earth, ordering them in a graduated scale of complexity. Nothing in this model necessarily implied that they hadn't all been created at the same time.

But Linnaeus's depiction of humans as a part of nature more or less on a par with other species, rather than distinct from and superior to them, seemed at odds with the biblical version of creation. Hence, the controversy. Linnaeus himself, by all accounts, was a firm believer in biblical teachings. He considered his contributions to be a further revelation of God's work, rather than a challenge to Christian belief.[21] In retrospect, Linnaeus has earned a place among the founding figures of biology through his development of the binomial system of classification (genus, species), which became a lasting contribution to the science.

This was an era in which the study of natural history flourished, involving the attempt to classify as many forms of plants, animals, and minerals as possible. The term "race" came into more frequent use in that context, often referring to classes of animals and plants.

In its archaic usage the term "race" referred to the course of a running stream, and in the eighteenth and early nineteenth centuries, it maintained much of this dynamic sense. The French naturalist Georg Louis Leclerc Buffon, in sorting and classifying tens of thousands of life-forms, used the term to distinguish related or similar types, but he maintained the sense that these divisions were somewhat arbitrary and could change through time.[22] Buffon, a rival of Linnaeus, did not share the Swedish scholar's sense of reverence for divine creation but adopted a more firmly secular approach to the material.

The idea of fixed human races became more established by the mid-nineteenth century. Still, in this early period, most scholars who addressed the issue of human difference maintained a clear sense of the effects of environment and historical experience in shaping characteristics. The naturalist Johann Friedrich Blumenbach, to whom we owe the "racial" category *Caucasian*, believed that various "races" had developed regional differences after departing from the original form. This prototypical human type, he believed, had most likely resembled the people of the Caucasus region near the Black Sea, whom he believed were the most beautiful of all human populations.

This, of course, also implied that "race" was a changing phenomenon.[23] Blumenbach further held that there was no reason to consider one particular "race" superior or inferior to any other.[24]

In the early nineteenth century, the American physician Benjamin Rush, a signer of the Declaration of Independence, argued that the dark pigmentation of Africans was a disease resulting from the tropical environment in which they lived. He believed that given sufficient time in the more beneficial climate of North America, their skin would become lighter. He found support for this view in the case of an African man who apparently suffered from a rare disease that caused his skin to whiten in patches. Nowadays this might have been diagnosed as vitiligo, although it's impossible now, of course, to be certain.

Samuel Stanhope Smith, who would later become president of Princeton University, also argued against the intrinsic superiority of any particular "race." He argued that "racial" classifications, in any case, were purely arbitrary, since it would be possible to delineate any number of such divisions using a range of criteria. Like most of his contemporaries, Smith attributed many of the distinctive (or alleged) characteristics of particular populations to the effects of climate.

IT'S ALL UPHILL FROM HERE

A general model of regression or degeneration persisted well into the nineteenth century, although with the burgeoning of the Industrial Revolution, an assumption of human progress came to predominate. By midcentury, one of the best examples of this lay in the writings of Herbert Spencer, who gave us the dubious gift of social Darwinism.

By Spencer's time the science of biology was developing rapidly, much of it based on the assiduous work of many dedicated amateurs. In the previous century, as we've seen, there had been a quest to classify the natural world by such scholars as Linnaeus, Buffon, and Blumenbach. Much of their work had been supported by governments and other monied interests to promote the identification of new resources such as plants or minerals that might have commercial value.[25]

In England and in the United States, however, many amateur naturalists were wealthy gentlemen who had the leisure and wherewithal to assemble private collections of natural specimens. Dried plants, pin-mounted insects, shells of crustaceans, fossils, and the pelts of rodents testified to the tireless pursuits of these naturalists in the field (or their assistants). It would be some time before universities became the primary base of the sciences and the standing of "scientist" would come to require formal scholarly credentials. But it was an era in which avid quests for knowledge depended on observations of the natural world rather than, as in earlier centuries, on laborious poring over ancient texts.

As we've noted, this also coincided with a turning away from theological explanations of worldly events and phenomena, even if many of those with an interest in nature and its wonders retained their personal religious beliefs and affiliations. It was fertile ground for ideas of biological evolution, finding expression in the work of Charles Darwin, Alfred Russel Wallace, and others.

The general idea of a long process of evolutionary advancement, controversial as it was because of its inconsistency with the Bible, resonated nicely with the burgeoning social idea of the time that progress, as opposed to degeneration, characterized the overall course of human events. Progress carried a spirit of optimism. On the other hand, Herbert Spencer's idea of progress with competition as the driving force, which called for the predominance of the strong and elimination of the weak, also drew inspiration from Darwin's concept of natural selection. "Survival of the fittest" became a well-known phrase in the popular domain and remains a powerful concept under various names and slogans. Some see a family resemblance between the nineteenth-century expression of such ideas, laissez-faire economics, and the more recent economic philosophy we sometimes refer to as "neoliberal."

European views of populations in other parts of the world tended to be consistent with these ideas of progress through competition, adapting, one might say, to the intellectual environment of the era. Questions of human differences and curiosity about them remained, but now thinkers had a biological, evolutionary model to provide some answers. In an age when European colonial expansion over many parts of the world was well under way, concepts of progress through competition, coupled with the survival of the fittest as a sort of a natural law, corresponded nicely with expansionist and imperialist policies.

Religion did not disappear, of course. But in some respects it, too, adapted to the times. Missionary ventures in far-off places, spreading Christianity to heathens, became praiseworthy, even heroic, enterprises that warranted lavish contributions from organizations at home. Dr. David Livingstone, one of the most famous European missionaries of all time, was explicit about his wish to spread capitalistic enterprise to "Darkest Africa."[26]

The availability of a biological model also allowed versions of racism to acquire some of the forms that seem more familiar to modern ears. At the same time, theological debates continued to rage in the nineteenth century over supposed human differences. One of the most heated of these was whether there had been a single divine creation from which the various "races" had diverged (monogenesis), or whether there had been multiple creations in the first place (polygenesis).

Josiah Nott, a prominent physician, was among the most vocal of the polygenecists. Notwithstanding his rejection of a literal interpretation of Genesis, however, he had no doubts that the earth was only about six thousand years old. As he and his colleague George R. Gidden pointed out, paintings in ancient Egyptian tombs depicted individuals thousands of years ago who looked very much like modern Africans. This, they reasoned, would not have left enough time for the various "races" to have become so different.

Nott was quite explicit about "racial" differences.

In 1842, I published a short essay on *Hybridity*, the subject of which was, to show that the White Man and the Negro were distinct 'species;' " [punctuation and italics in the original.] In the essay alluded to . . . I maintained these propositions: —

1. That *mulattoes* are the shortest-lived of any class of the human race.
2. That *mulattoes* are intermediate in intelligence between the blacks and the whites.
3. That they are less capable of undergoing fatigue and hardship than either the blacks or the whites.
4. That the *mulatto*-women are particularly delicate, and subject to variety of chronic diseases. They are bad breeders, bad nurses, liable to abortions, and that their children generally die young.[27]

There's more in the same vein, but the gist of it is clear. As we'll see in the following pages and chapters, even the most outrageous assertions about humanity can nestle in carefully woven baskets of "long first-hand observation," "evidence," and even, if we're to believe the writer, "proof."

As the nineteenth century wore on, particularly following the Civil War in the United States, polygenesis faded from the discourse. Increasingly, explanations of purported human differences based on biology—often no less imaginative than the theological arguments—came to occupy center stage.

ROMANTIC SUPREMACY

Even after Darwinian evolution gained broad acceptance, however, perceptions of human differences sometimes took other trajectories. The so-called Anglo-Saxon movement offers one example. The basic premise, of course, was the innate superiority of the "Anglo-Saxon race." But this perspective owed more to the Romantic movement of the early nineteenth century than it did to science or evolutionary theories.

The Romantic movement arose in the late eighteenth century as a reaction to the rise of science and rationalism. It celebrated the dramatic in nature, the emotional in human experience, the subjective aspects of life through art, music, and literature. It was the era that produced the poetry of Keats, Byron, and Shelley; the dark fantasies of Edgar Allan Poe; the novels *Frankenstein* by Mary Wollstonecraft Shelley and *The Scarlet Letter* by Nathaniel Hawthorne. *Frankenstein*, in particular, offered a cautionary tale about the hazards of scientific tampering with the forces of nature. In music, it was the era of the masterpieces of Beethoven and Mozart, the passionate and tragic romantic operas of Giuseppe Verdi, and the thunderous works of Richard Wagner based on Teutonic mythology. In art, the demonic paintings of Francisco Goya and the works of Eugène Delacroix convey a deep sense of dark emotion that characterizes much of the art of the times.

These developments in some respects tended toward mysticism—particularly with regard to historical mythology and inherited (and inherent) ethnic qualities and destiny. Proponents of Anglo-Saxon "exceptionalism," we might say, looked back to glorious ancient roots involving conquests, deeds of valor, and special virtues lacking in other "races."

Not surprisingly, this appealed to many in the United States who considered themselves to be of Anglo-Saxon stock, and who saw the influx of other populations as troubling, at best. Unlike some of the eugenicists of later decades, however, these self-identified Anglo-Saxons tended to hold an exuberant optimism about the future. Being superior, after all, Anglo-Saxons were destined to expand in numbers and sooner or later would come to dominate most of the world.[28]

Many self-defined American Anglo-Saxons saw a natural affinity that linked them to the English and Germans. This view lost some traction, however, when World War I broke out and German armies under Kaiser Wilhelm quickly became "Huns" rather than distant cousins who shared a blonde, heroic heritage. Eventually, the term *Aryan* gained ground as an alternative in some circles, although it, too, lost favor with the rise of Nazi ideology and the appearance of the *Übermensch*.

By the first few decades of the twentieth century, in many circles Anglo-Saxon dreams of glory had either faded to wistful daydreaming or had shifted from optimistic grandeur to fearful distress at the growing numbers of the wrong sorts of immigrants pouring into the "Land of the Free."

PHILOSOPHICAL BIOLOGY

By the late nineteenth century, we can see the beginnings of an enterprise that has thrived and continued to this day. Briefly, it amounts to claiming the legitimacy of science to justify or validate what are little more than positions of social philosophy or political opinions. We could see this development of a scientific posture as protective coloration, a camouflage that claims the imprimatur of science for what fundamentally are subjective arguments. As we'll see, this has continued unabated into the present.

It's one thing to assert that some people are inferior because, for some reason, one finds them distasteful. But how much more convincing it is to invoke "scientific evidence" resulting from "objective research" to validate such a position. It's no longer just a matter of one's personal opinion (or, God forbid, one's bias or prejudice). No, the data don't lie. With such a posture, one can promote oppressive policies more in sorrow than in anger.

Under this general rubric, we've seen proponents of slavery and colonialism argue that such practices are only following natural law. (Even in 2012, at least one public figure advanced the retrospective view that slavery had been beneficial to African slaves, since it exposed them to civilization and helped them advance from a lower to higher stage. Such arguments were far more common, of course, in the nineteenth century South.) We've seen persistent assertions of male dominance as a biological imperative, though such expressions often employ alternative vocabulary. We've seen policies designed to prevent designated groups or categories of people from reproducing, even to the extent of imposing government-sanctioned sterilization without their consent or even their knowledge. We'll discuss all of these and other misuses of biology in later chapters.

At this point, though, we must emphasize that this in no way amounts to a critique of biology as a science or of the validity of the concept of evolution. It is, rather, a critique of pseudo-biology, of false biology, of the misuse, distortion, and misinterpretation of valid biology to push arguments of sociopolitical philosophy.

As we'll see throughout the rest of this book, a number of warning flags often pop up in the course of this saga. Typically, after an author has worked to impress the reader with a lengthy discourse on research findings or scientific conclusions on some fairly specific issue, we find such phrases as "therefore it seems clear that," "it seems almost certain that," "it would be surprising if this did not also apply to," and so on. The author signals—inadvertently, perhaps—that "I am about to leap from solid scientific findings across a gaping chasm to some social and political opinion on the other side, and I invite you to join me." Let the reader beware.

Biologically themed "just so" stories have changed fashion over the years, often when certain issues, such as the struggle for civil rights, the Vietnam War, or gender equality initiatives, become national issues. Generally, biologically themed explanations or models have worked to reaffirm or justify the status quo. If we understand conservatism as a position that change is unlikely or even undesirable, then these models have been conservative. The idea that humans are simply "doing what comes naturally" carries with it the implication that to do otherwise would *not* be natural.

We'll discuss individual versions of this theme in later chapters. For now, we might simply note that over past decades, for well over a century, biologically themed arguments have asserted the necessity, inevitability, or "naturalness" of slavery, of colonialism, of the inferiority of people designated a particular "race," of aggressive interpersonal behavior, of male dominance over females, of selfishness, and of war based on territorial "imperatives."

The most intense debates over these various issues have been associated with historic periods leading up to the Civil War, the war itself, the Jim Crow era of institutionalized racial discrimination, the civil rights struggles of the 1960s, the Vietnam War, and the struggles for women's rights. In general, arguments for biological determinism have been mobilized in periods when social unrest and attempts to bring about change have become critical matters of public concern.

BIOLOGICAL STORIES

Many models have drawn on analogies between propensities observed in human behavior and in other animals (although as we shall see, assertions based on the alleged behavior of "Stone Age hunters" have also filled the same purpose). The animal analogy introduces another issue that warrants further attention in a later chapter: the anthropomorphizing of animal behavior.

Thousands of years ago, Aesop did this to great effect in his moralizing fables. Who can forget the hare and the tortoise, or the fox and the sour grapes? But obviously this has limitations. Certainly, animals often remind us of humans in general, and some perhaps even lead us to think of some personal acquaintances in particular. Who needs to be reminded that humans often behave like animals, whatever we might mean by that? But when we rouse ourselves from such musings, shake our heads, and blink back to the realm of common sense, it's obvious that in many ways, humans display some crucial differences from other species.

Humans learn and remember a lot more than other creatures, and what we learn introduces far more variety into our behavior. We do have deep-seated biological imperatives, of course. No human can help learning a language, for example, unless he or she suffers a disability that prevents it. Yet as we know, humans speak thousands of different languages[29]—many more than that, if we count languages that have disappeared—all of which, by definition, are mutually unintelligible.

In the United States, we no longer have plantation slavery—at least not legally. Yet just a few generations ago many humans clearly demonstrated their capacity to enslave other humans, and tens of thousands of defenders of slavery died in the effort to maintain that practice. Was this a genetic imperative? Probably not. It would be difficult to find many outspoken defenders of slavery today. Yet no gene pool could change that rapidly in three or four generations. Obviously, the practice of slavery was a behavioral *possibility*, and it could not have occurred without some biological capacity to develop and maintain it. But certainly it was not a biologically driven instinct.

The struggle for civil rights still has very far to go, but things have changed in that regard since the 1960s. Here, too, unfortunately, we can see all too clearly that trends could easily follow a number of directions in the future, whether advancing, turning back, or going in some other trajectory. But whatever might happen in the future, its causes will be far more complex than underlying human biological drives.

Women still earn less pay, on average, than men, but this too has shown some signs of improvement over the past several decades. As of 2012, few women had positions in the highest levels of government in the United States, although many more women won election to the legislative branch that November. According to a survey that year, this country had ranked below more than seventy other nations with regard to the number of women participating in the upper levels of government, just above Turkmenistan. Is this a matter of biological imperatives? If so, is our biology so different from that of those seventy other countries? Or are far more complex, nonbiological factors at work?

The Vietnam War has long been over—although we could argue that subsequent wars have made no more sense than that one did. We could debate at length the extent to which the invasion of Iraq, for example, was more a manifestation of instinctive "territoriality" than one of a deadly combination of duplicity, avarice, and stupidity on the part of national leaders. Still, if one wishes to include dreams of world domination as an example of territoriality run amok, we could probably keep that discussion going.

But let's continue our historical perusal.

Chapter 2

EUGENICS: BAD PEOPLE, GOOD PEOPLE, BETTER PEOPLE

IN THE EARLY YEARS OF THE UNITED STATES, THE GENERAL makeup and origins of the national population had already begun a long series of transformations that continue to the present day. When Europeans first began to invade the continent on a large scale in the 1500s—mostly French to the north, Spanish to the south, and English in between along the Atlantic seaboard—native populations vastly outnumbered them. Vikings had settled on the North Atlantic Coast centuries before, but their settlements never lasted long enough for them to become permanent occupiers, unlike other European contingents by the end of the seventeenth century.

We have tantalizingly few accounts of native peoples' attitudes toward these newcomers during the earliest phases. No doubt they varied a good deal. We do get a sense that the Wendat ("Huron") north of the upper Great Lakes were not overly impressed by these relatively short, hairy men whose personal hygiene was not up to local standards. The Wendat amused themselves at the expense of Jesuit priests who were trying to learn their language by teaching them obscene phrases and sending them off to try them out in the community.[1]

Relations between European and indigenous peoples in the ensuing centuries were mostly destructive for the latter, but by the end of the Revolutionary War, Native Americans still composed a significant population in the continent. The new United States Constitution, ratified in 1788, recognizes "Indian tribes" as independent political entities who had inhabited the region long before the colonies existed. Subsequent Supreme Court decisions established that this recognition gives

Congress, rather than the states, the sole authority to deal with them—unless Congress explicitly gives the states specific powers, as it did in the Major Crimes Act of 1949.[2]

A decision written by Supreme Court Chief Justice John Marshall in 1831 redefined the status of Native Americans in the United States through the use of the phrase "domestic dependent nations." This preserved Native Americans' independence from state jurisdiction, but the "dependent" aspect clearly had important implications for their future. By the 1830s, the general, nonindigenous population of the Americas had already changed a great deal through immigration, and native peoples—particularly those east of the Mississippi River—had come under increasing duress.

Although the story may be apocryphal, Benjamin Franklin supposedly wrote in the mid-1700s about the influx of foreigners who spoke little or no English, who had little understanding of American values or respect for them, and who threatened to undermine the American way of life. He was referring to German immigrants. Many other Europeans were arriving as well. In the mid-nineteenth century, the potato famine in Ireland impelled tens of thousands to migrate to North America, where the local inhabitants (or their recent ancestors) had themselves been immigrants, but who now viewed these newcomers as barely human.

Throughout the following decades, as the population of the United States continued to grow, its constituent populations remained in flux. One of the few regularities, perhaps, was the tendency of those whose forebears had immigrated a generation or two earlier to be suspicious of newcomers to their adopted homeland, and even alarmed by the threats they perceived in their increasing numbers.

The German immigrants of Ben Franklin's time, many of them farmers seeking arable land unavailable to them in Europe, often settled in rural enclaves. Irish immigrants had left an impoverished, conquered homeland that had long been under oppressive English rule. More often they migrated to the larger cities of the Northeast. The English had all but destroyed the Irish clan structure, outlawed the bards who had transmitted ancient lore, and in general, had attempted to eradicate indigenous Irish traditions, including the language.[3] By the time of the potato famine in the mid-nineteenth century when disease destroyed most of the crop, many thousands of Irish were already on the verge of starvation.

Over the decades of the late nineteenth and early twentieth centuries, other events in Europe triggered more population movements. Slavic-speaking southern and eastern Europeans, Italians, more Irish and Germans, and a variety of others found their way to North America. More than 20 million immigrants arrived between 1880 and 1920, 8 million of them in the first decade of the twentieth century.[4] The need for labor on the West Coast also brought thousands of Chinese workers who, with Irish immigrants, laid the tracks for the transcontinental railroad. And they all, of course, entered a nation in which hundreds of thousands of people whose roots lay in Africa had been enslaved since the 1600s. Many of their ancestors had been in the colonies centuries earlier, long before the arrival of most Europeans or Asians.

In every era, people at the top of the socioeconomic order viewed those others, the newcomers, with misgivings. Their labor was essential. But when they became too visible, too numerous, or appeared to threaten the social order, the elites didn't hesitate to define them as a problem.

THE RIGHT SORTS OF PEOPLE

A major and perhaps fundamental aspect of class difference and other social boundaries rested on perceived economic interests. A small segment of the total population controlled much more wealth than all the rest put together, and many social devices worked to reinforce this. At times these measures involved heavy-handed, sometimes violent reinforcement: excluding some categories of people from particular areas, neighborhoods, or places of business, for example.

Much of this, as we know, rested on obvious characteristics such as skin color. But skin color had little to do with why Irish or other categories of Europeans "need not apply." Other discriminatory criteria included "style," manner of speaking, level of education, religion, shared values, class background, knowing which fork to use, and other such virtues. These criteria not only reinforced elite boundaries but also were themselves the products of an elite, economically advantaged up-bringing. Interesting variants were snobbish responses to the nouveau riche, who had sprung from the wrong class or ethnic background and somehow obtained wealth but who lacked the refined manner of people with "old money." Such a theme has provided rich material for literature over the years.

The refined qualities of the established elites seldom included the virtue of manual labor. Thomas Jefferson acquired a classic image as an avid gardener, a gifted builder, and an artistic landscaper, but he seldom, if ever, rolled up his linen sleeves and grabbed a shovel. Like other plantation owners, Jefferson relied on slaves for that sort of thing. In other times and places where slavery was not an option, the "great unwashed" army of laborers did the heavy lifting—often for twelve to sixteen hours a day, six or seven days a week, before labor unions finally fought successfully, with many casualties, for improvements in working conditions. The famous captains of industry who ran things, meanwhile, could spend their workdays—comfortably, we assume—in tailored suits, neckties, and polished boots.

Laborers obviously were essential for production and to generate wealth for the elites. But laborers could be troublesome. It was important to have enough workers available. It was even better to have more willing workers available than were needed. A "surplus army of labor" allowed employers to adopt a "take it or leave it" stance with regard to wages and working conditions, since there were always plenty of others desperate to step into the jobs.[5]

On the other hand, as more and more immigrants flooded into cities, particularly along the East Coast, unemployed and often desperate masses could become disruptive. Unable to find paid employment, many of them sought alternative ways to survive, sometimes outside the law. Densely packed cities held enclaves of Irish and eastern and southern European refugees, and typically, these pockets became associated with high crime rates. To many concerned observers, such urban areas appeared to be cauldrons of unrest, a threat to civilized society.

Some immigrants and former slaves who had jobs nonetheless posed a threat to the established order, especially as they began to organize in unions. The strike of Pullman porters in the late nineteenth century resulted in the government calling in federal troops to quash it.[6] In the so-called Haymarket riots in Chicago in the early twentieth century, gunfire broke out, resulting in the deaths of several workers and police.

In the South during the days of slavery and later during Reconstruction, a common social nightmare was that, given the chance, slaves or ex-slaves would run rampant. Apparently, the idea that slaves might feel that they had some cause for revenge had occurred to many of their oppressors. Thomas Jefferson, reflecting on the issue, poignantly wrote, "I tremble for my country when I reflect that God is just."[7] The image of rapacious black men ravaging white southern belles persists to this day in many quarters, and over the years it has resulted in hundreds of lynchings. (After multiple attempts, Congress never could muster the votes to pass a federal law against lynching—which meant that it never became a federal offense. In 2005, Congress finally passed a bill apologizing for this shameful record of failure, although the vote was not unanimous.)

We can see one major theme in all of this and possibly its central irony. The oppressors have generally seen themselves as potential victims, vulnerable, somehow, to those they oppress. On the one hand, masses of people bore the burden of labor, if they had jobs at all, and suffered numerous forms of exclusion and discrimination. On the other hand, a small elite benefited from the vast wealth that the masses produced for them, oversaw their oppression, and wielded all of the political power. Yet the powerful have typically viewed the oppressed as a menace.

So we can perceive the problem of the elites, whether or not we sympathize with it. And as we might expect, some of the deep thinkers of the time pondered possible solutions.

WHAT'S TO BE DONE?

To resolve a problem, of course, one needs to identify its root causes. Nowadays, students of the situation might look to historic, social, economic, or political factors. But at the time, that could have led to the uncomfortable conclusion that these troublesome masses were victims of circumstances they had not created. Some intellectuals, of course, did arrive at such a conclusion, and some became activists in trying to improve matters—whether through social action or charity work. But this required certain egalitarian assumptions; that is, the suffering masses were not fundamentally different from anyone else. They just had different histories and faced different circumstances.

For most scientifically oriented thinkers of the time, though, the root of the problem lay with the masses themselves. Many intellectuals believed that these troublesome people were very different indeed, at a basic, innate level.

We recall that by the late nineteenth and early twentieth centuries, biological explanations for perceived human differences had become an important intellectual theme. It's not surprising, then, that many intellectuals attributed the troublesome characteristics of these foreigners to inherent biological shortcomings. An important aspect of this idea was that, for practical purposes, their traits were fixed and unchangeable. They might be the result of long evolutionary processes, but the ponderous course of evolution based on natural selection was long and slow, and change would require many generations.

The renowned white supremacist Lothrop Stoddard expressed this view clearly. "We know that men are not, and never will be, equal. We know that environment and education can develop only what heredity brings." His greatest nightmare, apparently, was that "the white race . . . will be swamped by the triumphant colored races."[8] The famous botanist Luther Burbank referred to the lowest echelons of society as "human weeds." "Nature eliminates the weeds, but we turn them into parasites and allow them to reproduce."[9] David Starr Jordan, president of the University of Indiana, wrote in 1902 that "the pauper is the victim of heredity, but neither Nature nor Society recognizes that as an excuse for his existence."[10] Jordan later moved on to become president of Stanford University.

For living people, it was too late for evolutionary change to be of any use. They were what they were. To many it seemed that the only hope for the future of humankind was that more of the fittest should survive and produce larger numbers of their equally fit offspring. The living mass of undesirables could not undergo change through any well-meaning but unscientific social programs intended to help them, since their characteristics were fixed, "in the blood." Long-term biological change might offer some hope, but only in the distant future. The most immediate concern was not to improve future generations of undesirable immigrants, but to help save the wider society from their onslaught in the present. It was common in the eugenics literature to find references to the poor or disabled as "bugs and pests."[11]

It seemed imperative that measures to slow or even to stop immigration should begin at once. As we know, this concern has remained foremost in many circles even now, when numerous politicians continue to fret about the United States being overrun by foreigners—in this case, from Mexico and Central America rather than China, Ireland, or Southern or Eastern Europe as in the past. Many of these anti-immigrant politicians are only one or two generations away from their own immigrant ancestors. But to return our attention to the early twentieth century: Even if it were possible to stop immigration, what about those immigrants who were already here?

John Harvey Kellogg, now better known as the inventor of corn flakes, had interests that extended far beyond breakfast cereal. In 1914, he sponsored the First Race Betterment Conference in Battle Creek, Michigan. The leading eugenicist Harry Laughlin pretty much laid it on the table. "The compulsory sterilization of certain degenerates is therefore designed as a eugenical agency complementary to the segregation of the socially unfit classes, and to the control of the immigration of those who carry defective germ-plasm."[12]

As if the unruly ways of these foreigners weren't bad enough, immigrant masses clearly were reproducing faster than the established elites. The reasons for this might seem puzzling, but in England, the eminent sociologist Herbert Spencer had offered an explanation. A renowned proponent of progress through competition, Spencer explained that even within the human body, cells of various types compete with one another.[13] Within that competitive arena, brain cells compete with reproductive cells. Thus, among those who employ their brain cells to the fullest, the procreative cells tend to atrophy. Conversely, those who place fewer strains on their intellect and let their brain cells take it easy can invest more energy in relatively mindless reproduction. Hence, according to Spencer, the lower classes are bound to outstrip the elites in producing all of those ill-mannered, squalling brats.

Spencer's point might seem a bit silly to many modern readers and justifiably so. As we'll see in chapter 4, however, in biological determinist tracts a century and a half later, J. Philippe Rushton still found it to be a compelling idea.

We now know, of course, that the rate of reproduction among the world's populations shows a close relationship to the rate of infant mortality. Impoverished masses have many children because so many of them die. Such a principle might have raised uncomfortable questions about improving living conditions of the poor to help decrease the birthrate. But in Spencer's time, such a proposition would have seemed absurd.

The most desirable approach for some would have been to encourage the elites, the most "fit," to reproduce more. The other side of the coin was to find some way to stifle the reproductive rates of those who were *less* fit. Even if it weren't feasible to compel these elites to get busy and have more superior babies, a more practical solution might be to find a means of decreasing the reproductive rate of the lower classes, which ultimately would have the same effect.

Short of attempting to promote sexual abstinence among the lower classes—good luck with that—one obvious way was to sterilize the less fit. That way they could behave or misbehave as much as they wanted to no effect. It might be difficult to find volunteers for such a remedy, but society provided ways for enlightened policy makers to implement the process on an involuntary basis. Prisons and public hospitals offered plentiful opportunities for such procedures. Diagnoses of deficiency that could justify such action were easy enough to produce. Why, after all, would people be in prisons or asylums in the first place if they weren't unfit?

PARASITES AND PESTS

Many eugenicists saw charity as a serious mistake. Franklin Bobbitt pointed out that charities interfere with survival of the fittest.[14] Crusader Margaret Sanger, founder of the birth control movement and an avid eugenicist, wrote of the "cruelty of charity," which perpetuated "this dead weight of human waste."[15] In the *Birth Control Review*, a eugenicist wrote that "it is interesting to note that there is no hesitation to interfere with the course of nature when we desire to eliminate or prevent a superfluity of rodents, insects or other pests; but when it comes to the elimination of the immeasurably more dangerous human pest, we blindly adhere to the inconsistent dogmatic doctrine that man has a perfect right to control all nature with the exception of himself."[16]

The criteria for fitness arose fundamentally from economic strata, but to some sensitive ears, such a statement might have sounded overly crass. This was, after all, a democracy. But to many, poverty itself was an indicator of deficient fitness. Experts developed a range of indicators of the need for remedial action. One of the favorites was "feeblemindedness." Whatever guidelines one might use to diagnose such an affliction, most medical authorities considered such a trait, like many others, to be biologically innate and therefore not subject to therapeutic correction or cure. Other such innate characteristics included criminality, promiscuity, and, as we have noted, a propensity to be poor.

In 1911, a eugenics committee met in New York as a section of the American Breeders' Association with the financial support of the Carnegie Foundation and Mary H. Harriman, representative of one of New York's most distinguished families. They compiled a precise list of the types of "unfit" specimens who might warrant eugenic attention. In addition to paupers and the feebleminded, these included alcoholics, criminals, epileptics, insane people, those who were "constitutionally weak," those who were deemed especially susceptible to particular diseases, those who were deformed, and those who were deaf, blind, or unable to speak. Edwin Black notes that "in this last category, there was no indication of how severe the defect need be to qualify; no distinction was made between blurry vision or bad hearing and outright blindness or deafness."[17]

There were critics. New York psychiatrist Smith Ely Jelliffe, who happened to be a good friend of the arch-eugenicist Charles Davenport, objected to the broad use of the term *insanity*. Insanity, he pointed out, could have many causes, ranging from head injury to various medical conditions.[18] In 1927, a prominent professor of biometry at Johns Hopkins characterized eugenics as "a mingled mess of ill-grounded and uncritical sociology, economics, anthropology, and politics, full of emotional appeals to class and race prejudice, solemnly put forth as science, and unfortunately accepted as such by the general public."[19]

As the twentieth century progressed, geneticists took pains to distinguish their valid scientific research from the eugenics movement. Such critiques did little to dampen the zeal to improve the human race—or at least, a select part of it—by coercing some of the rest. It would take the Holocaust to do that.

As Edwin Black observed, "Eugenics was nothing more than an alliance between biological racism and mighty American power, position, and wealth against the most vulnerable, the most marginal, and the least empowered in the nation."[20] The *Journal of the American Medical Association* was very much on board. Reporting on the International Eugenics Congress in London in 1912, the esteemed medical journal gushed with praise. "Its correspondent enthusiastically portrayed the eugenicists' theory of social Darwinism, spotlighting the destructive quality of charity and stressing the value of disease to the natural order."[21]

Strivers for a better society through eugenics were not necessarily marginal cranks or malcontents. They had the support of some of the most powerful institutions and wealthiest individuals of the day.

Much of this movement, of course, subsumed the more general issues of "race" and ethnicity—two ideas that for many at the time were coterminous. (Such terms as "the Jewish race," "the Slavic race," "the Nordic race," and so on, were common.) In these times, well before the age of political correctness, arguments for the inferiority of entire groups generally were far from subtle, although they did vary somewhat. In the view of some, people of African descent were beyond improvement, despite the achievements of such distinguished public figures as Frederick Douglass or W. E. B. Du Bois. It was easy enough to dismiss such cases as aberrant. Native Americans—especially those who had lost their lands, resource base, and cohesive communities and found themselves among the lower strata of urban populations or on impoverished reservations—also occupied an inferior social status. Irish, eastern Europeans, and southern Europeans shared a difficult common existence, often grudgingly, at the lower levels of society.

The eugenics movement in the early twentieth century gained energy celebrating superior types, or "stocks," from northern Europe. The inspiration for the term *stock* arose from the selective breeding of cattle and other domestic animals to enhance desirable qualities. This Nordic movement in the 1920s attracted interest in many intellectual drawing rooms and inspired writings that remain an embarrassment to a number of scholarly disciplines to this day. Several scholars wrote books extolling these gifted descendants of the Vikings and Teutonic tribes and their mythical valor.[22] Their works also attracted attention in Europe, where the young Adolph Hitler found them intriguing and very much in line with his own thinking. From prison in the 1920s, he wrote to the American eugenicist Madison Grant stating that Grant's book *The Passing of the Great Race* was "his Bible."[23]

Some of the leading figures in the eugenics movement in the 1930s found encouragement, and even vindication of their work, in the Nazi efforts to eliminate the unfit. Generally, this approbation predated the news of concentration camps and the Holocaust. But Charles Davenport, head of the Cold Spring Harbor eugenics database, happily shared data with Nazi experimenters in the 1930s and applauded their efforts.[24] His colleague Harry Laughlin visited Europe in that decade to offer his advice.

Most eugenicists on the European continent had rejected American eugenic policies of forced sterilization and incarceration of those deemed unfit, considering them violations of basic human rights. The Nazis, however, saw the American approach as a model for their own methods.[25]

Today, the Nordic movement has lost much of its social luster and tends to appear more frequently in marginal social circles: neo-Nazi and skinhead groups, for example.[26] As with many bad ideas, though, it would be unrealistic to assume that it's disappeared entirely even in mainstream thinking.

In the late nineteenth and early twentieth centuries, proponents of these and related views had a far more receptive audience. It was obvious that the disorderly throngs of the lower classes presented a problem. Involuntary sterilization seemed to be a humane solution. It seemed best to prevent these sorts from passing on their inferior genes to future generations, who, after all, could hope to be no better than their progenitors.

MANIPULATING BIOLOGY

Much of this thinking, as we've seen, fell under the general rubric of "eugenics," which means something like "good breeding"—an idea promoted by Sir Francis Galton, a statistician who happened to be a first cousin of Charles Darwin. Many celebrities and other prominent figures of the era, such as George Bernard Shaw, Theodore Roosevelt, H. G. Wells, Woodrow Wilson, and Winston Churchill, heartily endorsed it. Margaret Sanger, heroic figure in the struggle for women's rights, was all for it. Notwithstanding her concerns for justice, she strongly advocated forced sterilization and incarceration of the unfit—at least during their reproductive years.[27] Such a policy, in their view, would address a serious problem that many of the less fit failed to admit or even to realize they had. If society could solve the problem for them without their even being aware of the blessing, how could it be wrong? Where was the downside?

By the 1930s, more than thirty states had passed mandatory sterilization laws for numerous categories of people who came to the attention of the authorities.[28] Many of these individuals did not have drastic disabilities or show propensities for violence. Many were people the appropriate authorities judged to be feebleminded or perhaps just were poor. By 1935, various government agencies had authorized the sterilization of about thirty thousand people, often without their knowledge or consent.[29]

Leaving aside instances of botched procedures, many of these patients were none the wiser. We can never know how many Native American women entered government hospitals for various health reasons and were sterilized without their knowledge or how many inmates in prisons or in asylums were sterilized. In some cases, eugenics specialists advocated sterilization not only of schizophrenics, but of their families as well.

Armed with the rediscovery of Gregor Mendel's work on inheritance of traits among peas (but with a crude understanding of genetics, at best) some specialists advocated sterilization of the families of those with a range of deficiencies. Seizing on the idea of recessive traits, they believed that even though individuals in these families might show no flawed characteristics, they were liable to be carriers of undesirable traits that could appear in future generations.

It's impossible to know how many women went to their graves without ever understanding why they had never been able to have children. We do know, however, that in Vermont, native Abenaki patients were subject to the procedure well into the 1970s.[30] Not that Vermont necessarily stands out in this regard. More likely, they simply kept better records than many other states. In some impoverished rural Appalachian communities, families lived in dread of government workers arriving at their houses to take children in for "treatment."[31] Involuntary sterilization of women continued well into the twentieth century in California. In 2012, the North Carolina issued a formal apology for having sterilized at least 7,600 people from 1929 through 1974, either without their knowledge or against their will.[32]

Some of what took place under the term *eugenics* had a far more benevolent aspect, especially in England, where the term originated. By the end of the first decade of the twentieth century, Galton and most of his colleagues had parted ways with the extreme, coercive American views. What Galton preferred to call "positive eugenics" could amount to such measures as improving prenatal care. Who could object to that? And some observers who felt concerns for the betterment of humankind (and who doesn't?) saw a kind of inadvertent eugenics taking place spontaneously before their eyes, as better public transportation allowed increased contact among distant communities, thus reducing the amount of inbreeding in small, formerly isolated towns.[33]

Further along the sliding ethical scale, many supported the policy of encouraging certain "desirable" sectors of the population to reproduce through benign, positive measures. Some advocated measures to reduce the frequency of inherited diseases such as sickle cell anemia, Tay-Sachs, and others.

By the late twentieth century, however, researchers better understood that genes for many of these problems are, in fact, recessive alleles—diseases show up only if the individual is homozygous for the trait, carrying both alleles. Otherwise, the presence of a single allele in an individual may actually confer benefits. With sickle cell, the presence of one allele confers heightened resistance to malaria. Most inherited diseases are not genetically simple—relying on the presence or absence of a single gene as the sickle cell trait does—but depend on the interactions of multiple genes. These can sometimes number in the hundreds, not to mention the epigenetic, environmental interactions involved in such diseases appearing in an individual.

Early in the twentieth century, however, great and perhaps naïve optimism prevailed in many quarters with zeal to improve humanity at large. Eugenics became an academic discipline in which many universities, including Harvard, Yale, Princeton, and Syracuse, offered courses. In many respected academic institutions, eugenics achieved the status of a respected, legitimate, even scientific field of study.

Some writers see the continuation of a form of eugenics in modern medical practices such as prenatal screening and "designer babies." Whatever one might think of genetically designed babies, however, today's practices differ from the eugenic practices of the bad old days. For one thing, they're voluntary: matters of choice for the (adult) individuals involved. These choices, in the United States at least, are often available only to a financially advantaged sector of society.

Earlier eugenic programs, for the most part, were not matters of personal choice and tended to be directed at particular targeted and vulnerable groups or at individuals within these groups rather than at elites. Policies aimed to prevent birth rather than to encourage it. One horrific exception was the Nazi selection of blonde, "fit" women in occupied countries to be raped by German troops and forced to bear the rapists' "superior" children.

As we've seen, multiple themes wove through these ideas. Aside from "race" and such alleged inborn traits as promiscuity, criminality, or a predilection for poverty, "intelligence" played a large role. Feeblemindedness, as we have seen, was a common diagnosis, but such a diagnosis was susceptible to distortion by all sorts of subjective factors, ranging from language problems to personal antipathy. By the 1920s, however, proponents of "racial improvement" had acquired a new tool: a test measuring an individual's "intelligence quotient," or IQ.

THE GIFT OF IQ

The IQ test did not originate as a means of sorting out those who were intellectually inferior from those who were superior in terms of their innate abilities. Alfred Binet, the French psychologist who first devised the test, had intended the IQ test to identify children who needed remedial help in school to catch up to their peers.[34] The basis of his idea was that these students not only *could* be helped, but *should* be helped.

Henry H. Goddard and Lewis Terman of Stanford University, American specialists in education, began to apply it differently.[35] They began with the assumption—by that time essentially an article of faith—that low performance on IQ tests was a result of innate, fixed biological traits. Thus, the American Stanford-Binet test in many cases became a device for identifying those who *could not* be helped and who, therefore, were destined to fall by the wayside. It became a tool for weeding out the less fit, determining winners and losers at an early age, and it also became a useful weapon for eugenicists. This gift to education would play an important part in "tracking," separating supposedly "gifted" students for special treatment in public schools, which began in the mid-twentieth century and persists in many school systems today.

Armed with the IQ test as a supposedly scientific, objective means of gauging intellectual fitness and bolstered by the assumption that such traits are immutable, eugenicists went to work. The idea that scores on such tests might reflect other issues such as cultural and linguistic background, education, home environment, and general life experience was largely brushed off the table like so many annoying crumbs.

A few awkward anomalies cropped up, however. Researchers in the 1920s had noted that, in general, African Americans (though this was not the term used at the time) scored more poorly on the tests than "whites" did. No surprise there. Researchers also noted that African Americans in the North did better, on average, than those in the South. Could this reflect a difference in education, in economics, or in other experiential or environmental factors? Not likely, according to the thinking of the time. Since everyone knew that intelligence is linked to "race," it must be that "Negroes" in the North had more "white" ancestry mixed into their backgrounds than those who had remained in the South.

In the 1920s, the army decided to apply IQ tests to troops on a broad scale, testing both "white" and "Negro" troops by administering the so-called Alpha tests. It was easy to separate the results for the two categories. Integration would have to wait for another quarter century when President Harry S. Truman desegregated the armed forces. But in some respects the Alpha results made no sense. "White" troops from the North did better than "Negroes" from the North, and much better than "Negroes" from the South. Oddly, northern "Negroes" also scored higher than southern "Whites."[36]

How could this be? It was hard to stretch the "white admixture" explanation to account for these strange results. Surely, African Americans in the North couldn't have more "white blood" than southern "whites" . . . could they? As we noted, the idea that the test scores reflected education and life experience—that significant causal factors were aspects of environment rather than biology and were amenable to change rather than innate—didn't get much of a hearing among proponents of the "scientific" measurement of intelligence.

A number of prominent voices did raise objections, however. As early as the 1920s, as the IQ testing enterprise refined its tools toward more mathematical precision, the pioneering geneticist Lancelot Hogben warned of "the danger of concealing assumptions which have no factual basis behind an impressive façade of flawless algebra."[37] Addressing a wider audience, the outspoken columnist Walter Lippmann pointed out that "intelligence is not an abstraction like length and weight; it is an exceedingly complicated notion which nobody has as yet succeeded in defining."[38] In a rather testy exchange with Lewis Terman, Lippmann wrote, "I hate the impudence of a claim that in fifty minutes you can judge and classify a human being's predestined fitness in life. . . . I hate the abuse of the scientific method that it involves. I hate the sense of superiority which it creates, and the sense of inferiority which it imposes."[39]

BOAS AND "FIXED TRAITS"

The idea of immutable inherited characteristics faced another challenge early in the century through the work of Franz Boas, a figure whom many consider the "father of American anthropology." Boas was interested in immigrant populations entering the United States from Europe. He himself was an immigrant from Germany, and his training in physical geography at Heidelberg and the University at Kiel had instilled rigorous scientific methods in his work.

Boas was an inductivist. He was an old school researcher, who believed that scientists should collect data without prior assumptions and then draw conclusions from the evidence. Most researchers nowadays use a more deductive approach, forming hypotheses that they test against the data. If the available data fail to disprove the hypothesis, then it remains plausible—for the time being, at least. In reality, most research involves a blend of the two. It would be difficult to come up with a hypothesis without having considered pre-existing data, which seems fundamentally inductive in nature. Nor would any strict inductivist cling to conclusions that later information failed to support.[40]

In Boas's time, many of those in positions of authority were concerned about the health, fitness, and general qualities of new immigrants. In 1910, the U.S. government commissioned Boas, already a prominent scientist, to look into the matter. Inductivist that he was, Boas set about collecting massive data on new immigrants and their children, including those born in the United States. He gathered thousands of measurements of various characteristics, such as stature, body weight, head size, and even head shape. The cranial measurements turned out to provide some of his most intriguing data.

As we might expect, Boas found that head shape and the size of the skull varied considerably among immigrants from different geographic regions. More surprisingly, however, he also found that the shape of the head—the cephalic index, which expresses the ratio of cranial length to width—differed significantly, and consistently, between adult immigrants born in Europe and their own children born in the United States.[41]

Here was a clear indication of the profound effects of environment on human growth. It was (and still is) difficult to imagine why head shape mattered in terms of adaptation, or even more puzzling, how environmental conditions could affect head shape in one generation. But surely that was precisely what had happened. There was no conceivable way that this change in physical aspect (phenotype) could have been a result of any change in the genes (genotype); not in such a patterned, consistent way within two generations. What had happened, clearly, was a matter of inherited genes expressing themselves differently in different environmental contexts. If something as concrete and tangible as the shape of the skull could turn out to be so malleable, how could anyone be certain that complex behavioral patterns were inherited as immutable traits?

Boas's conclusions excited those who were already skeptical of the "fixed trait" school of thought. But it seems to have had little impact beyond a narrow scientific audience, and even there it met with skepticism.[42] Its influence on government policy, in the heyday of eugenics and national xenophobia about the influx of foreigners, seems to have been almost imperceptible.

Later in his career, Boas would become a vocal and influential public figure on many social issues, particularly civil rights. But in this era, as in so many others, scientific findings were not enough to dissuade true believers. This was just a few years before the federal government began rounding up recent immigrants as suspected radicals in the Palmer Raids, and before the Johnson-Reed Act of 1924 restricted immigration from Russia, the Balkans, and other parts of southern and eastern Europe.[43]

It probably would have been too much to expect a wide spectrum of the public to alter deeply held beliefs because some stodgy scientist, and an immigrant Jew, no less, had found some surprising statistical results in a mass of his own data. Many Americans, even in the twenty-first century, remain unpersuaded by the mass of scientific evidence for evolution. Many prefer instead to hold the idea that God created the world and everything in it about six thousand years ago in only six days of work, taking the next day off. Boas's findings appeared at a time of eugenic fervor with a strong dash of xenophobic attitudes toward immigration—scientific findings be damned.

But a revisiting of Boas's work in the last decade or so also brought a disheartening response. Conscientious scientist that he was, Boas had taken steps to ensure that the raw data he collected would be available to future researchers. Almost a century later, two researchers did, in fact, reexamine Boas's data and claimed to have discredited his conclusions.[44] They did so by pointing out that the intergenerational differences in head shape that Boas had found were, in fact, quite small. As a result, some of the mass media, including the *New York Times* proclaimed that Boas got it wrong.[45]

But as others quickly pointed out, the new research proved no such thing. Boas at the outset had pointed out that the differences in cranial indices were small, and he gave figures to demonstrate the comparisons. The crucial issue was that consistent differences existed *at all*. As one anthropologist pointed out, the latest researchers had actually confirmed Boas's findings.[46] The media greeted this important but less exciting news with a bored silence. As Lee D. Baker observes, the public who were even aware of the issue probably still believe that modern researchers refuted Boas's findings.[47]

OUT WITH EUGENICS, FOR NOW; BUT WE STILL HAVE IQ!

The heyday of eugenics lasted into the 1940s, but it became a bit of an embarrassment with the rise of Nazism in Germany, as news of the Holocaust eventually reached the American public. This is not to say that Hitler lacked fans in the United States, including a few of the most famous names of the times. But for many Americans, by the end of World War II, the Nazi example had demonstrated the logical and horrifying extremes of eugenic approaches to social issues. Eugenics did not disappear altogether, but its proponents felt the need to pipe down a bit.

For many, on the other hand, intelligence tests continued to be useful tools. Alfred Binet's relatively benign purposes, to identify students in need of remedial work, largely fell by the wayside. But the basic idea of testing gave rise to a lively academic industry.[48] Under the leadership of Lewis Terman and others at Stanford, IQ testing came to focus on identifying the academically "gifted" as opposed to those who were, well, not so much. In this incarnation, IQ testing took on the sterile, polished-chrome semblance of a scientific endeavor. Since the testing enterprise carried the aura of dispassionate precision, only a few critics seriously debated what the term *intelligence* meant. That discussion, which could include such imprecise, nonquantitative issues as creativity, social skills, or any of a substantial number of other qualities, seemed unsuitable, somehow, for scientific, objective measurement.

We might note that such qualitative factors as creativity or social skills may be much more complicated and difficult to measure than standard academic subjects, especially with a quantitative metric. The scientists of intelligence, however, chose the more simplistic option. Intelligence was what the intelligence tests measured. And what they measured was the ability to answer questions "correctly," as judged by the test constructors. What the tests measured was intelligence because they were *intelligence tests*. Duh!

The evident circularity of this argument made some observers a bit uneasy, and even some proponents acknowledged that there were flaws in those early attempts. But those glitches, no doubt, would be corrected as researchers continued to refine the techniques and methodology. This was "science," after all, and science *progresses*.

As time passed, though, some critics were impertinent enough to continue raising questions about whether the test scores might be affected by such things as life experience, including social class, quality of education, or some interaction between the two. We had, after all, the strange case of northern "Negroes" outscoring southern "whites" in the Army test early in the twentieth century. And late in the twentieth century, a researcher named James R. Flynn pointed out that in the previous few decades average IQ scores had been rising in many countries.[49] Hmmm. If IQ is inherited, or "fixed," how to account for the "Flynn effect"? Had some sort of rapid evolutionary change occurred? That would be weird. Still, in 2013, a writer could still note that "many psychologists tell us that intelligence is an enduring individual trait, pretty much hard-wired by a person's DNA and by cell development in the fetus's brain."[50]

Critics' concerns have not cramped the expansion of the thriving testing industry. In fact, in retrospect we might view this as a golden age for that enterprise. For one thing, test scores are so convenient! And precise! They actually provide a numerical scale for ease of comparison, dispelling any sense of subjectivity. Of course, a good teacher, if he or she were interested in gauging a child's abilities or potential, would certainly consider multiple aspects of a child's background, personality, fears, aspirations, and so on—all of the things, in other words, we subsume within the concept of "person." That's all well and good, but such an exercise is difficult and costly. It's inefficient, both in terms of time and emotional energy. It certainly is handy to have those IQ scores. IQ scores have continued to play an important part in American life, long after the heyday of the eugenics movement.

As we know, the debate continues. We can clearly see the influence of testing in the No Child Left Behind Act of 2001 and its consequences, charter schools, outcome assessments, Race to the Top, the Common Core mandates for public schools, and so on. The testing industry, producing standardized, machine-graded exams by the millions, is doing quite well, thank you very much.

Eugenics didn't die out completely. In the 1970s, it came to the fore in the works of such exemplary figures as William Shockley, a winner of the Nobel Prize in Physics, who will receive more attention in the next chapter. "Racial" differences in IQ test scores also became an issue again in the 1960s and beyond, when, as we shall see in chapter 4, Columbia University psychologist Arthur Jensen, the political scientist

Charles Murray, and a number of others had their moments of fame or notoriety, depending on one's point of view. In the 1960s and beyond, the focus in the United States would shift more directly to "race," particularly African Americans as contrasted with "whites," rather than European immigrants as it had earlier. In the first decade of this century, scholars such as Steven Pinker of Harvard, a recent darling of the media, also have taken up the cause of asserting innate differences in intellectual abilities. But that's a discussion for a later chapter.

Chapter 3

KILLER APES, NAKED APES, AND JUST PLAIN NASTY PEOPLE

WORLD WAR II AND WIDESPREAD REVULSION AT THE horrors of Nazism dampened enthusiasm for biological explanations of human differences, at least for a time. It never disappeared entirely, and for some students of human nature, assumptions of innate differences among human populations continued as a more subdued, underlying theme. Unabashed overtly racist arguments were distinctly less acceptable in the public domain.

By the 1960s, some writers had found a different focus for a biological theme. In this case it was to broaden the scope to ideas about human nature in general, rather than focusing on alleged differences among categories of people. The issue became more a matter of what we're *all* like as a species, whether we might want to admit it or not. Undercurrents of "race" eventually did reappear in the works of some of these writers— almost like an underground stream that occasionally reveals itself in low spots on the surface.

The 1960s in the United States was a time of great political turmoil and demands for change. Major historic figures were assassinated, including President John F. Kennedy, Robert Kennedy, Martin Luther King Jr., and Malcolm X. The war in Vietnam gave rise to massive protests. In 1970, National Guard troops at Kent State University in Ohio shot and killed students demonstrating against the war on their own campus.

Popular feelings in the country ranged from calls for radical change to demands for harsh measures to maintain "law and order." It was a time when forces for reform and forces to defend the status quo stood in stark and violent opposition.

By that time, science in various fields had, of course, advanced greatly since the 1930s. In addition to advances in technology, our knowledge about the human past, and particularly the course of human evolution, had become much clearer. These developments offered rich material for writers who sought examples from biology to expound their views.

ENTER THE KILLER APE . . .

By the late 1960s, exciting discoveries in human paleontology had occurred in southern and eastern Africa. As early as the 1920s, a researcher named Raymond Dart had come across a fossilized skull, including the face and an internal cast of the cranial cavity that showed characteristics of both apes and humans. The braincase was small, comparable to that of a chimpanzee. But unlike chimps, this creature had small canine teeth. Dart named the animal *Australopithecus africanus*, which we could translate as "southern ape of Africa."

In subsequent years, a number of comparable forms turned up in East Africa, especially through the work of Louis and Mary Leakey and later, their son Richard. Eventually, a picture emerged of two or more species of hominins[1] that had once inhabited a broad range from southern to northeastern Africa, perhaps as early as four million years ago. These creatures had walked upright and had small canine teeth comparable to those of modern humans, rather than apes. Much of the picture remained unclear, however. By the twenty-first century, researchers had identified numerous lines of australopithecines and other early hominin forms and pushed the time back millions of years.

In the 1960s, before this rich array of finds became available, though, *Australopithecus africanus* pretty much held the spotlight. The relatively empty canvas invited a great deal of creative speculation, and one who accepted the invitation with enthusiasm was a playwright named Robert Ardrey. Ardrey's previous accomplishments had included the screenplay for the movie *Khartoum*, starring Charlton Heston. (There is no apparent connection between this and Heston's later starring role in *Planet of the Apes*.) Ardrey became fascinated with what these African fossils could tell us about the human experience—not just our past, but our present.

Ardrey speculated that if we could understand more about these australopithecines, we could have a better grasp of our own heritage. And by "heritage," Ardrey had in mind a sort of inheritance that we've kept with us to the present, rather than having left it by the evolutionary wayside. One of these traits, Ardrey claimed, was a propensity to kill. Australopithecus, in his view, had essentially been a "killer ape."[2]

Before we consider the evidence he presented for this assertion, we might ask why, even if this had been true of Australopithecus, it should also be true of us. Needless to say, a million years is a long time, especially if we estimate a human generation to be about twenty years or so. We're talking about something like 50,000 generations—and probably many more, since australopithecines probably had a shorter life span than modern humans. If we grant that a significant portion of human behavior is a result of learning, then we have to assume that a tremendous about of learning must have taken place over those tens of thousands of parent-child cycles and life spans.

But herein lies a key aspect of the issue. If we were to assume that some behavioral traits are biologically innate and pass down genetically through generations, then the importance of learning as a source of behavioral malleability might not be that great in some respects. A behavioral trait persisting for millions of years—tens of thousands of generations—may not be implausible. If physical traits such as small canine teeth could appear early and persist for so long, why not other characteristics? The argument would of course rest on the assertion, once again, that a significant portion of behavior is genetically "built in" rather than acquired through learning and experience and that it can be passed on, largely unchanged, over an immense span of time.

Even if this were the case, though, what made Ardrey think that little *Australopithecus africanus*, standing only about four feet tall with wimpy teeth, was such a killer? Ironically, the teeth had a lot to do with it.

Australopithecus lacked the physical equipment we might expect of any decent predator: no claws, no fangs, and probably the slowest running speed of almost any other mammal on the savannah. Even in modern times when we have longer legs, a human would be hard-pressed to catch a skittish housecat. But *Australopithecus* possessed an advantage that no housecat, let alone a wild carnivore, could ever have. Australo-pithecines had chipped stone tools, and probably heavy sticks or long bones for clubbing or throwing at prey. According to Ardrey, they were able to hunt and kill game because they were capable of creating and using weapons. He bolstered his case by pointing to the discovery of a cave containing the skulls of baboons with signs of breakage from violent blows.[3]

Ardrey also claimed that this australopithecine killing instinct went far beyond hunting. He pointed to remains of australopithecines that showed signs of violence. The bottom line was that in the course of hunting for meat, our alleged ancestor had evolved to become a killer ape that not only preyed on other species but often killed its own kind. The nasty little creature apparently had australopithecidal tendencies.

To evaluate this, we need to unpack the model a bit. For starters, we might consider whether *Australopithecus* was, in fact, as committed to hunting as Ardrey suggests. Other researchers have offered alternative hypotheses: that *Australopithecus africanus* was mainly a scavenger, more prey than predator,[4] or that it was mainly a seedeater.[5]

Let's suppose that australopithecines *did* hunt and kill other animals for food. What would that imply about their alleged murderous proclivities? Do other predatory species often kill each other?

Not usually. It does happen; but generally, other species have means of controlling conflict or keeping it below lethal levels. Wolves' surrender posture usually ends a fight. In historic times, the few human populations who have lived by hunting had fairly peaceful reputations. Homicides do happen all over, of course. But if we're looking for people who kill other people on a regular basis, we really need to look at agricultural societies or state systems with much larger populations. People who live by hunting game and gathering wild plants tend to have small, scattered populations. They can't afford to lose many people, especially productive adults in their prime who bring in the food. Life is tough enough as it is. Cannibalism occurs here and there, but again, we have far more cases of cannibalism among settled agricultural peoples than among hunting societies.

When Ardrey conflates killing animals for food with killing members of one's own species, he seems to argue that the same motivations, drives, or instincts underlie these forms of behavior. But what are some of the most common reasons why people kill one another, aside from war? Personal hatred? Responses to insult? Fits of rage? Sexual jealousy? To steal a victim's property?

How many of these motivations might apply to hunters killing game animals?

Most observers would agree that a hunter killing an animal is not usually angry with it or driven by hatred. Most likely, he and his family are just hungry. And to suggest any of the other motivations—insult, jealousy, property—would be silly.

In many hunting societies, the etiquette of the hunt requires the hunter to show respect for the animals. Native American hunters in the subarctic considered the relationship between the hunter and the game to be cooperative, which meant that the animal chooses to present itself to the hunter. This relationship would be lost if the hunter insulted the animal's spirit in some way.[6] This is a far cry from the rampages of a "killer ape."

The killer ape model also poses some logical problems from an evolutionary perspective. In the face of selective pressures that confront all species through the course of time, what evolutionary advantage could there be for members of a species to have a propensity to kill one another? Ardrey addresses this by introducing another element to the argument: an instinct to defend territory, what he refers to as the "territorial imperative."[7]

According to that model, the evolutionary advantage would not have been a simple, primordial Hobbesian war of all against all. Instead, it would have been a case of one local group facing off against other groups of intruders. From that proposition, it was a short step to arguing that those groups who were most successful at defending their territories thrived, while others who were less successful fell by the evolutionary wayside. Thus, Ardrey deftly shows us how killing members of one's own species might, in fact, have evolutionary advantages.

One problem with this—aside from the lack of any direct evidence that such a propensity ever characterized australopithecines—is that among living or historically known human populations, territorial boundaries—especially those requiring defense—are far more common among agricultural and state-level societies. Most hunting peoples had a population density of less than one person per square mile and needed to be mobile because of a scattered, shifting food supply. There's no reason to think that australopithecines ever exceeded or even approached historic human population densities. One small hunting band in the western subarctic of North America exploited a territory of about five thousand square miles.[8] Defense of far-ranging territorial boundaries wouldn't have been feasible even if people had been inclined to do so—whether "instinctively" or not.

Ardrey's tracts appeared when the United States was engaged in the Vietnam War, an enterprise that, to a growing number of people, seemed irrational if not downright criminal. Many who objected to the war took their protests to the streets and other public arenas. But the general population was polarized on the issue, with many other people outraged that the protesters questioned government policies. In that climate, Ardrey's views on the supposedly innate, natural basis for aggression and territorial conflict found a receptive audience. Those who are old enough may remember the "domino theory": "if we don't stop them over there, we'll have to do it over here." From a territorial imperative perspective, that seems to make a lot of sense.

. . . FOLLOWED BY THE NAKED APE . . .

In the early 1970s, Desmond Morris, a primatologist and an Oxford don, advanced another argument to support an evolved, inherent basis for much of human behavior. Morris had more scholarly credentials than Ardrey but no less imagination. Morris proposed to show a continuum between humans and our nearest relatives, the living apes and the ancestors we share with them.

One of the most obvious differences between us and our close relatives, of course, is that we no longer have an effective coat of thick hair over most of our bodies. (We actually do have about as much body hair as a chimpanzee, but most of ours has become finer and transparent.) Morris, presumably in an attempt to emphasize our relationship with our simian cousins, dubbed us, as a species, the "Naked Ape."[9] He drew parallels between various human reflexes and behavior patterns with those of the "other" apes. Morris considered the human smile, for example, to be analogous to the "fear grimace" of a chimpanzee.

Unlike Ardrey, Morris's work shows more interest in sex than in violence. It makes for lively, entertaining reading. He addresses the puzzling fact that human female breasts remain permanently enlarged after puberty, unlike those of our primate cousins which enlarge only during lactation. Human female breasts also enlarge somewhat during lactation, but as we know, they don't disappear altogether at other times. Morris's explanation (whether ingenious or not, the reader may judge) was that enlarged female breasts actually play the role of displaced buttocks. That's right. Buttocks.

Among other primates, he explains, the sexual approach is from behind. Sexually receptive females of many primate species, when sexually receptive, make their situation known through the appearance of their rear ends in sometimes flamboyant and colorful ways. These displays, needless to say, are of great interest to males.

Fascinating as all of this is, we might consider a few problems. For one thing, among the higher primates only humans—we naked ones—have buttocks. The twin masses of muscle in that area, primarily the *glutei maximi*, developed as an aspect of our upright posture. A proud achievement of human evolution, they allow us to stand upright for extended periods of time. Apes, though they obviously do have rear ends, have never developed this feature. They rarely stand erect for any length of time.

Presumably, if human female breasts were in fact simulations of the rear ends of apes or other primates, they probably ought to have been flat, calloused, and subject to various colorful displays at certain times.

Which brings us more directly into the realm of human sexuality. Unlike many species, humans have no particular breeding season. Human births occur in every month of the year. This does not mean, of course, that individual humans may not feel more or less receptive at various times. But as we know from novels, movies, TV sitcoms, and possibly even experience, humans have other, and perhaps even more complex, ways of communicating this receptivity to prospective sexual partners. Finally, from his model, Morris assumes that the "normal," standard human sexual approach is from the front. O-kaaay. No need to mention the *Kama Sutra*; that would be awkward.

. . . FOLLOWED BY NASTY PEOPLE

Another enterprise linking human behavior to our biology arose in the 1970s in the work of Konrad Lorenz. Lorenz, a distinguished scientist, had won a Nobel Prize for his work with greylag geese and fish. He and his colleague Nikolaas "Niko" Tinbergen had studied imprinting, the process in which goose hatchlings become attached to the first creature they see after breaking out of their shells. Normally, of course, this would be their mother. In his research, Lorenz managed to have little geese imprint on him instead. In classic, iconic photographs we can see Lorenz, a bearded, portly gentleman, strolling along at the head of a line of devoted goslings.

Lorenz's contributions to the science of ethology—animal behavior— were significant and widely influential. Later in his career, however, when he turned his attention to human behavior, his findings were more controversial.[10]

Like many writers we've already discussed and others who would later hearken to the biological muse, Lorenz began with an assumption of biological determinism. He asserted that "man's whole system of innate activities and reactions is phylogenetically so constructed, so 'calculated' by evolution, as to *need* to be complemented by cultural tradition [italics in the original]."[11] Okay, so we do have culture, but it's only because our biology *makes* us have it. This might sound somewhat innocuous, but does that mean that the "whole system of innate activities and reactions" *drives* or *shapes* culture in some way? It all depends on how far this "innateness" extends in human activity. Sneezing, okay. Sex drive, fine. In Lorenz's view, however, it goes far beyond that. Lorenz is willing to concede some importance to culture; yet collective, learned behavior (or in other words, *culture*) becomes a direct product of genetic directives.

We can see here an early version of the tendency to attribute anthropomorphic qualities to evolution, which Lorenz claims "calculates" the need for culture. It's hard to imagine how calculation can take place without some sort of intentionality. Lorenz and other biological determinists would emphatically deny that they are making such attributions, of course; that would be odd. It could even sound mystical, with some superhuman entity manipulating things. Very unscientific. The use of quotes signals a *metaphor* (wink, wink)—shorthand for a process that's far too complicated for them to explain to us right now. Except that they never do. The specific evidence for just how this happens is apparently misfiled somewhere. Maybe it just got lost.

Of course, we could argue that the evolutionary process has resulted in a species that depends on *learning* to survive, a species that's developed a higher capacity for flexibility and, hence, adaptability, because of a *diminution* of biological scripting of behavior. But that would relegate biological imperatives to the background rather than center stage, where these writers are "determined" to keep them.

The "metaphorical" attribution of intentionality to the evolutionary process, which, as we shall see, appears continually in the writings of biological determinists, despite the lack of supporting data, comes to resemble a bizarre sort of secular doctrine of faith. But at the center of their devotion is not simply a belief in evolution per se, but a social philosophy that certain modes of modern behavior are natural—almost sacrosanct—because they've evolved. What this implies about the prospects for progressive social change is clear.

In the later work of Lorenz, we can see the posture of the "generic expert," the idea that expertise in one field implies an aura of unbounded expertise in others, especially those pertaining to human behavior. One could say that it blurs the distinction between personal brilliance and actually knowing what one is talking about. We get the sense that, for many of those who have immersed themselves in "hard science," general human behavior seems so easy to understand that anyone who is bright enough can analyze it without necessarily making a serious study of it.

Ardrey had no formal scientific credentials, although his abilities as a writer were well established. Morris was, in fact, a trained primatologist and knew a great deal about the behavior of apes. He was even a consultant for the opening "chimp" scene for *2001: A Space Odyssey.* In Morris's controversial later work, he extrapolated from ape to human behavior and more or less equated the two, at a level that to many seemed far beyond what the data warranted.

Lorenz's work with greylag geese and fish did not provide him with any special background to analyze human behavior. He was, of course, a human being, and on that score had much to say about the human experience, but no more than anyone else. Yet he did feel that his studies about biologically driven behavior in birds had much to tell us about ourselves. No doubt he was right to a degree—but to a limited degree. One is reminded of the old saying that when all you have is a hammer, everything looks like a nail.

The sort of behavior that most interested Lorenz was, in effect, misbehavior, in the sense that he focused on aggressive tendencies among humans. Like Ardrey, Lorenz saw innate aggressive impulses as a part of the human evolutionary heritage. In his discussions he seems to have males, particularly, in mind. He touches on aggressive behavior at both the individual and collective levels, but asserts that even "communal aggression" (war, for example) is "clearly distinct and yet functionally related to the more primitive forms of petty individual aggression."[12]

In discussing the experience of "militant enthusiasm," he really gets into it. When a person finds himself in such a state, according to Lorenz, "A shiver runs down the back and . . . along the outside of both arms. One soars elated, above all the ties of everyday life, one is ready to abandon all for the call of what, in the moment of this specific emotion, seems to be a sacred duty."[13] Unfortunately, as he points out, "Men may enjoy the feeling of absolute righteousness even while they commit atrocities."[14]

At a physiological level, "the tone of the entire striated musculature is raised, the carriage is stiffened, the arms are raised from the sides and slightly rotated inward so that the elbows point outward. The head is proudly raised, the chin stuck out, and the facial muscles mime the 'hero face,' familiar from the films."[15] The "hero face," he asserts, has its counterpart in the male chimpanzee.[16]

Despite the difficulties of imagining the "hero face" with jutting chin on a chinless chimpanzee, the picture Lorenz offers us seems oddly anachronistic and, in a rather chilling way, historically familiar. We can't help but wonder whether this posturing is a "vestige of a pre-human vegetative response,"[17] as he asserts, or merely a vague recollection of another historic time and place in the author's own experience.

The aggressive potential of humans seems beyond doubt. On any given day, the daily newspaper has plenty of examples, but the causes of such behavior are less clear. Many social scientists, philosophers, and psychologists have spent their careers exploring these issues. For Lorenz, though, the answer was far simpler. Human beings, he asserted, are inherently aggressive and prone to violence. "Like the triumph ceremony of the graylag geese, the militant enthusiasm in man is a true autonomous instinct; it has its own appetitive behavior, its own releasing mechanism, and, like the sexual urge or any other strong instinct, it engenders a specific feeling of intense satisfaction."[18]

For Lorenz, the question was not so much why we act aggressively, but what inhibits us from doing so even more. "Humanity is not enthusiastically combative because it is split into political parties, but it is divided into opposing camps because this is the adequate stimulus situation to arouse militant enthusiasm in a satisfying manner."[19] One could almost visualize a vehicle in gear with the engine racing, ready to spring forward the instant the brake is released.

IT GETS WORSE

As controversial as Ardrey's and Lorenz's ideas were, overtly racist arguments did not dominate most of their writings. Their assertions, in general, applied to humankind at large. Fundamentally, they both felt that we all have an innate tendency to be pretty nasty. But the undercurrent was apparent, especially in the case of Lorenz.

During his days as a young Austrian scientist in the 1930s, Lorenz had affiliated himself with the Nazis. In his earlier writings, as he admitted later, he had expressed beliefs in a superior "race" (guess which one) and had concerns that the best qualities of the elite were threatened by admixture with inferior types. He had been a eugenicist in his beliefs. Later in life he expressed regret for those earlier views and essentially renounced them.

But even in later years, Lorenz's perspective was far from egalitarian. His writings left little doubt that he was convinced of crucial differences among various "racial" groups, including innate, biologically fixed differences in intelligence. He based his model on his work with greylag geese.

Lorenz and Tinbergen had observed that the behavior patterns and responses of wild geese changed when they were domesticated. Because they no longer faced the challenges their wild relatives had to confront, they tended to spend more of the energies in eating and having sex. They tended to "lose their edge," one might say. Lorenz felt that the same process had been occurring among humans as a result of civilization. Many of the old instincts (and, yes, he did use the term "instincts" in discussing modern humans[20]) had tended to atrophy because they were no longer crucial for survival.

One aspect of this, Lorenz felt, was that members of a particular "race" might no longer feel as strong an aesthetic preference for their own kind, a preference that in precivilized times, he believed, had maintained sexual boundaries between populations. As a result, an unhealthy tendency to mate across "racial boundaries" had developed. The superior "race," therefore, was bound to suffer in quality from this laxity that came with civilization. Lorenz had no apparent doubts that the "races" were not equivalent, particularly regarding intelligence.

A SERIOUS FLAW IN THE ARGUMENT

There are many things wrong with this, of course, but to begin, we might examine a basic assumption that Lorenz and many others have held: that "race" is a valid scientific concept. A near consensus has prevailed for many years now among the relevant scientific community (that is, those who have actually researched and studied the subject) that "race" as a bounded entity is not a meaningful scientific construct, since human populations with distinct genetic boundaries do not exist.[21] For this reason, the word "race" appears here in quotes.

Genetic diversity exists throughout humanity. But when we look at the actual distribution of *real* genes, it's clear that genes cannot define or delineate "racial" groupings, or "races," with any sort of genetic boundaries. Human genes occur in a type of distribution that researchers refer to as "clinal."

This means that a given gene might occur with a very high frequency in a particular geographic locality. As we move farther and farther from that locality, fewer people have that particular gene, although, even at a great distance, it might still be present in a few individuals. The frequency, or rate, of occurrence of a particular gene typically decreases with distance, although there may be other centers of high frequency for the same gene in other places. A given gene, in other words, may appear in multiple clines. Such a distribution can result from either local selective pressures for the particular gene or gene flow, genetic transmission among neighboring people.

Another crucial aspect of the clinal distribution of genes is that clines for *different* genes almost never coincide precisely. If we were to draw clinal maps of the distribution of even a few genes, they would overlap in what appears to be a chaotic pattern (or, rather, the lack of any apparent pattern). Imagine what a map would look like if it portrayed the clinal distribution of the entire human genome, which consists of thousands and thousands of genes (about twenty thousand at last estimate). Any two random individuals within a single population are likely to have more genetic differences between them than occur in the collective genetic frequencies between any two populations.[22]

A favorite counterargument to assert "racial" difference is to point to specific genes that occur in one place but are rare, or even absent, in other regions. The so-called sickle cell gene is a common example. Some have suggested that the sickle cell gene is an obvious marker of African ancestry.

The sickle cell gene is a genetic response to selective pressures resulting from the presence of falciparum malaria, a particularly deadly form. Those who are heterozygous for the sickle cell allele—that is, they have one allele for the sickle cell trait and one allele that doesn't carry the trait—have heightened resistance to malaria. But those who are homozygous for the sickle cell allele, having two alleles for the trait, generally die at an early age. Those who lack an allele for the sickle cell trait but have two normal alleles at that locus escape the problem of the sickle-cell disease but are more susceptible to malaria.[23] This, of course, would be a serious disadvantage in areas where the disease is especially prevalent.

Thus, in regions where this deadly form of malaria is common, counteracting selective pressures have maintained a fairly steady frequency of the gene. In nonmalarial regions, the disadvantage of the sickle cell trait has no counteracting advantages, and selective pressures tend to eliminate it. Rather than being a "racial" marker, the sickle cell trait is a malaria-zone marker. Similar blood adaptations occur in other zones far outside of Africa that also have high rates of malaria: southern India and parts of the Mediterranean region, for example.

We might also note that even where the sickle cell trait occurs, it appears only in a minority of the population—about 17 percent in West Africa, for example. If the sickle cell trait were a "racial" marker, would that mean that the 83 percent or so of the population without the gene belong to a different "race"?

All discussion of "race," therefore, ultimately rests on a concept that has no basis in empirical evidence. "Race" is not real except as an imaginary entity, quality, identity, or socially constructed category.

I KNOW, BUT IT STILL SEEMS REAL

This is far more than an attempt to prove a negative—or the nonexistence of something—which many coffee-table philosophers hasten to tell us is impossible. Rather, positive evidence (or "proof," if you will) exists that "race" as a rational biological construct or bounded entity based on genetic distribution is a fallacy. What we now know about the actual frequencies of genes throughout the human population contradicts the old ideas of distinct "racial" groupings. They just aren't distributed that way.

Although genetic variations exist among populations, "race" doesn't exist in the real world. For it to exist as a biological phenomenon, a significant cluster of genes would have to be frequent in one population but absent, or nearly so, in all others. But few if any genes are unique to a given population. Any of the genes that might define "racial" group-ings are scattered across the globe. "Race" is a social myth disguised as a scientific category, imposed on real people who often have no more in common genetically than any two individuals selected at random from different parts of the world.

As Jonathan Marks points out with regard to the familiar ABO blood types, "a large sample of Germans . . . turn out to have virtually the same allele percentages (A = 29, B = 11, O = 60 as a large sample of New Guin-eans (A = 29, B = 10, O = 61). A study of Estonians in Eastern Europe (A = 26, B = 12, O = 57) finds them nearly identical to Japanese in eastern Asia (A = 28, B = 17, O = 55)."[24]

A common fallback argument is the "sure, *now!*" position.[25] Even though things might be a little mixed up nowadays, some maintain, there was an earlier time in human history when the "races" were more pure. It's only been in recent times, with more frequent interregional contacts, that more "racial" blending occurred—or so the argument goes.

It would be difficult to measure the degree to which genetic mixture among different populations has increased, but it's beyond doubt that from the age of early European colonial conquests through the present time of jet travel, contacts between people in widely separated parts of the globe have intensified. And when people get together, we know what often happens. Gene flow.

In recent years, Alan Templeton's work[26] has pretty much blown away the idea that "pure races" ever existed among ancestors of living human populations. Using what he called a "nearest neighbor" model, Templeton has shown that gene flow among populations has continued throughout human history, ultimately involving even the most distant groups in the process.

Obviously, before the era of jet travel, sexual interactions among people living ten thousand miles apart would have been rare and prob-ably even nonexistent. But interactions with one's nearest neighbors would be a different story. And just about everybody is a nearest neigh-bor to somebody else. As a result, few if any specific genes have re-mained exclusive to any particular population. Most genes are, liter-ally, all over the place, occurring in varying frequencies. That, as we recall, is what clines are all about. The living populations of humans don't stem from any formerly isolated, genetically distinct groups. We've always been part of a single global gene pool.

The Human Genome Project has stirred a much interest in genetic distributions and given rise to a lucrative industry purporting to trace people's ancestry—generally for a fee of at least a few hundred dollars. Unfortunately, ancestral results can be misleading. Simple kitchen-table math will show that a mere ten generations ago—or about two hundred years—each of us is a direct descendant of 992 different people. All of them, of course, have their own complex ancestries.

More important, perhaps, the clinal distribution of genes means that populations throughout the world differ almost entirely in the *frequencies* of genes, rather than having a specific repertoire of genes all their own. Particular genes—with the possible exception of a few rare mutated forms—are not very useful as ancestral markers. Some genes are rare in certain geographic areas and common in others. But ancestral searches inevitably involve a degree of guesswork—or perhaps we should say, estimates of probabilities. No doubt this is one reason why people who have used these services have often gotten different results from different searches.

The Human Genome Project has been extremely useful in giving us a better picture of our genetic makeup as a species. It has also greatly diminished the probable number of genes it contains, from more than one hundred thousand in earlier estimates to about twenty thousand. But it has done nothing to revive or substantiate the old idea of "racial" boundaries. If anything, it's underscored the assumption of constant gene flow among populations for a very long time.

We must allow, however, that just because something doesn't exist in concrete reality—in what we like to call the real world—doesn't mean that it can't exist in some other sense. Think of Santa Claus. Just because the jolly old elf may not really be working up at the North Pole all year making toys doesn't mean that he doesn't constitute a real and important aspect of American culture—with important real-world economic consequences. Unfortunately, the imaginary construct of "race" has been far less benign.

That's not to say that isolated groups and genetic divisions have never existed in the distant past. Perhaps as recently as twenty-five thousand years ago, different forms of *Homo sapiens*—Neanderthals and modern human forms—coexisted. Some Neanderthal genes seem to have contributed to our own genetic heritage. But the crucial issue is that neither Neanderthals nor any other human variants have survived into recent times. All living humans spring from a single human population sharing a common gene pool.

BACK TO EUGENICS

Konrad Lorenz's late-life mellowing, such as it was, did not apply to another Nobel laureate, William Shockley. Born in England in 1910, Shockley had a distinguished career as a physicist in the United States at Bell Laboratories, where he and two colleagues invented the transistor. They received the Nobel Prize in Physics for this contribution in 1960.

As formidable an achievement as this was, far fewer of the American public at the time would have become familiar with Shockley's name had it not been for his outspoken views on "race." He became increasingly vocal on the subject during the last decades of his life.

Shockley's views amounted to little more than old-fashioned eugenics. Like Francis Galton, Charles Davenport, and Harry Laughlin before him, Shockley feared for the future of humanity—or at least, what he considered to be his branch of it—because he was convinced that average human intelligence had been decreasing as a result of "racial" admixture. Shockley had no doubt that "Negroes" were less intelligent than "whites" or that this alleged characteristic was an innate, biologically inherited trait. One solution, he asserted, would be to offer these less fit people financial incentives to seek voluntary sterilization.

Yes, we've heard all this before. Shockley offered another example of social philosophy operating under the guise of scientific omniscience, adopting the protective camouflage of blanket expertise. His brilliance as a physicist seems beyond question. His brilliance as a social engineer, however, is something else altogether. In the 1980s, an *Atlanta Constitution* article characterized Shockley's views as blatantly racist. He sued for defamation. The judge ruled in his favor, but awarded him one dollar in damages.

Although racism, as we know, has never disappeared, its mode of expression has altered at various times. By the second half of the twentieth century, overt expressions of racism had become less acceptable in public discourse. Openly racist groups continue to exist in the United States in the twenty-first century, perhaps more commonly in some parts of the country than in others. After the election of President Barack Obama, the number of racist groups drastically increased. But often, even the most unabashed racists have felt a need to be a bit more discrete than in the past—calling themselves "White Citizens' Councils," for example, rather than the "Ku Klux Klan," or adopting still more benign-sounding names. Most racist epithets in the public realm have become a bit more circuitous—referring to Obama as "the Food Stamp President," for example, rather than some of the epithets people might have used in the past to make the same point.

We can see some parallels between the views of the young Konrad Lorenz and the old William Shockley, and in contrast, perhaps, more similarities between the views of the elder Konrad Lorenz and those of Robert Ardrey. One thing they all shared throughout was an assumption that observable behavior is generally a manifestation of inborn biological factors.

In chapter 4, we can see a few old, familiar ideas dusted off, given academic credentials, and put to use in response to civil rights initiatives in the 1960s and beyond. This time the issue had to do with whether society could, or should, improve the well-being of all of its members.

Chapter 4

MIND GAMES

THE ISSUE OF IQ TESTING, WHAT IT MIGHT REVEAL ABOUT various groups and individuals, and what policies it might justify simmered for decades after Alfred Binet first developed the concept. It finally boiled over in the 1960s. Many questions had remained over the years. What do these tests measure? Are they accurate assessments of overall intelligence, or are they tests of learned abilities? Do they predict success in life, or do they merely reflect the success that one's parents have already had in producing a nurturing environment?

One attempt to resolve disputes over the significance of the tests early in the twentieth century had come through the work of Charles Spearman. Spearman devised a statistical measurement that became known as Spearman's g, with the "g" standing for "general intelligence." At its basis, this amounted to an assertion that the tests did, indeed, reflect overall intellectual abilities. Many others have disagreed. Steven Jay Gould referred to it as a "mathematical artifact"—". . . the abstraction of intelligence as a single entity, its location in the brain, its quantification as one number for each individual, and the use of these numbers to rank people in a single series of worthiness, invariably to find that oppressed groups—races, classes, and sexes—are innately inferior and deserve their status."[1]

One of the most fundamental questions associated with the test was the extent to which it reflects innate abilities rather than environmental variables such as poverty, education, test-taking experience, or even the relationship between the person taking the test and the administrator. A number of studies have shown that these factors are indeed significant in affecting test outcomes.[2]

We can imagine that such factors as the test subjects' previous experience or lack of experience with tests, feelings of apprehension, a history of deprivation at home, chronic hunger, a sense of hostility or expectations of poor performance conveyed by test givers, taking the test in English as a second language, and so on, are likely to affect the outcome. As a comparable example, one can imagine that in a sporting contest, a beginner is not likely to do as well as a practiced, well-trained athlete, whatever the innate physical potential of either of them might be.

The issue of "racial" differences underlay many of these questions. On average, African American students have not scored as well on these tests as "white" students. One aspect of these findings is that for the most part, samples of the "white" students whose scores were taken into account in these studies have attended well-funded suburban schools, while the samples of African American students attended poorly funded urban schools. By the middle of the twentieth century, this represented the typical distribution of such populations. (People of a certain age may remember the phrase "white flight" pertaining to this demographic shift in the mid-twentieth century, when many "whites" moved out of cities into the suburbs as more African Americans, including many from the South, moved into northern urban areas.) Taken collectively, the qualities of schools in different areas and the educational benefits they have been able to offer students have not been equal.

One could easily posit a few "yes, but" examples. Elite schools, such as Boston Latin or the Bronx High School of Science, have certainly existed in urban settings; and many African Americans have lived outside urban areas. But considering the fact that public school budgets rely on local property taxes, many predominantly African American schools in nonurban areas have suffered by comparison because of smaller budgets—unless, of course, they happened to exist in wealthy predominantly African American suburban communities. Such cases have not been typical.

SOCIAL PROGRAMS? NOT SO FAST

Issues of educational disadvantage intensified in the 1960s when civil rights became a national concern. A growing number of citizens became more conscious of the extent of racial discrimination in this country, and public opinion grew in opposition to it. President Lyndon Johnson pushed civil rights legislation outlawing voter suppression on the basis of race. He also established a national program called Project Head Start. This program's aim was to overcome some of the disadvantages affecting preschool children in impoverished areas. Its purpose was to promote better preparation for success in school. Many of these children were African American, although a majority of them were not.

Many citizens, including academics, applauded this initiative. Others fretted about the costs. A few academics rose to object.

It would be difficult to make a case directly against the idea of helping children in need. An alternative position, however, was to argue that the efforts of President Johnson and other liberals, while perhaps benevolent and even laudable, were misguided. Why misguided? Because according to these experts, intelligence and academic achievement are only partially a result of environment. Generally, they're matters of heredity and can't be remedied by costly social programs. Such experts often presented their views in regretful tones; it was more in sorrow than in anger that they delivered this sad news.

Notwithstanding the conciliatory tone, though, this message involved more than a passing on of regrettable but undeniable findings. The thrust of the message was part of an effort to undermine and discredit what many conservatives considered to be leftist social engineering programs with unacceptable costs. The public perception that the main targets or beneficiaries of these expenditures would be African Americans was, beyond doubt, a part of the picture.

This assertion might seem unjustified, since most of the players who objected to the program were careful to avoid statements of that sort and to couch their objections in relatively diplomatic ways. But of course, the general issue of "racial" differences in intelligence had not arisen de novo. It was at least as old as the army tests of different "racial" groups half a century earlier in World War I, with far deeper roots than that. This later episode in the 1960s was, among other things, merely another chapter in a continuing age-old saga.

Arthur Jensen, a psychologist at Columbia University, was a main participant in the discussions. In 1969, Jensen published a paper in the *Harvard Educational Review* that addressed differences in IQ scores among African American and "white" students.[3] Jensen argued that because the major component of intelligence is inherited—genetically fixed—any attempt to change it through social or educational programs was doomed to failure. Jensen asserted that the genetic component of intelligence is about 80 percent.

We could pause for a moment and wonder why, even if this were true, it might not be worth it for a society to do its best to nurture and cultivate even the 20 percent of intelligence subject to environmental influence. After all—what might society be like if the overall level of intelligence (whatever that might be) increased by even a couple of percentage points? But that, of course, would be to attach a hypothetical speculation to what amounts to an unfounded assertion in the first place. It might be more useful to consider how Jensen arrived at his 80 percent. His major source was a British researcher named Sir Cyril Burt.

SEEING DOUBLE

Burt was renowned in the science community for his studies of identical twins. In theory, identical twins should provide vital insights into how much of any trait, including intelligence, is inherited. Identical twins begin life with the same genes. If they become separated soon after birth and grow up in different circumstances, it should be possible, from observing them later in life, to determine which characteristics they share because of genes and which differences they display because of different life experiences.

An improperly healed broken arm from playing high school football or a missing incisor from walking into a door in ninth grade fall at the easy end of this spectrum. So, too, do differences in height or musculature resulting from diet or exercise. Many other factors, such as speaking English rather than Urdu, are also easy to account for. But other issues may be less clear-cut.

Identical twins, since they were born at the same time in the same place, are likely to be subject to many of the same cultural factors associated with the period of time in which they live, particularly if they both remain members of the same society and occupy comparable economic strata. Factors of this sort could range from immersion in the popular culture of the times to prevailing political attitudes.

There's no way to control for all of this, of course. To raise individuals in completely isolated conditions to control for all external influences would be extremely difficult, ethically appalling, and, in fact, probably impossible. Human infants need a good deal of contact and interaction with other people to survive. Thus, whatever conclusions one might reach about twins raised separately in real life would necessarily rest on a certain amount of sloppy data. There have, nonetheless, been some rather striking, if not outlandish, cases.

In 1979, *People* magazine ran a story about two men in Ohio who were identical twins separated at the age of three weeks. They'd been unaware of each other's existence until they were thirty-nine years old. At that point in their lives, according to the report, they discovered that each of them drove a blue Chevrolet; as children, they each had had a dog named "Toy." And both of them had married women named Linda, divorced them, and subsequently married women named Betty.[4]

It's doubtful that even Jensen or Burt would suggest a genetic basis for naming one's dog "Toy" or some inborn attraction to women named Linda or Betty, let alone a biological imperative to marry them in that sequence. We can only continue to wonder about the strange tale of the twins in Ohio for now, although we will hear about them again.

The genetic factors that Burt and Jensen claimed were somewhat more pernicious, since they amounted to pronouncements about the potential of large numbers of individuals to achieve success in life. Jensen's "80 percent" figure was the basis for his argument against public programs designed to assist disadvantaged children. The origins of that figure lay in Cyril Burt's work.

WAIT—WHAT WERE THOSE SCORES AGAIN?

One problem with Burt's methodology was that his sample of adult identical twins who had been raised separately was very small. Not many more than a hundred recorded cases were known globally at the time and fewer still in either Britain or the United States (subsequent researchers have located many more, as we shall see later). The small sample size would make any broad generalizations based on these findings rather questionable. But a more serious issue arose after Burt's death in 1971. Researchers examining his work discovered that he had fudged his data.[5]

Burt was so eager to demonstrate what he apparently believed to be the truth, it seems, he tweaked the data to be sure they revealed the proper results. The scores of his identical twins were not only close; they were too close. They were closer, in fact, than scores the same person would be likely to achieve taking the test on two different occasions.

We could speculate on Burt's motives, but in a sense it really doesn't matter. The crucial factor is that these findings had a major influence on subsequent assertions about the inheritance of human abilities. And they were wrong.

The context in which Burt carried out his research in England was somewhat different from that in which Jensen wrote in the United States in the 1960s. Civil rights and "race" were in the forefront of discussion and subject to heated debate. Burt's focus was less on "race" than on social class, but the thrust of his conclusions was comparable. People who demonstrated less achievement, in his view, did so because of their own biologically inherited limitations rather than their inherited economic and social disadvantages. None of this mushy, Dickensian blaming of social conditions. No use trying to improve something that can't be changed.

Project Head Start went on as planned despite objections by those who argued that it was a waste of resources. A follow-up study published in December 2012, almost half a century after its inception, concluded that "at the end of 3rd grade, the most striking subgroup finding was related to children from high risk households. For this subgroup, children in the 3-year old cohort maintained sustained cognitive impacts across all the years from pre-K through 3rd grade."[6]

So there, Dr. Jensen.

YOU SAY HERITABILITY, I SAY INHERITANCE—
LET'S CALL THE WHOLE THING OFF

Much discussion of these issues, especially in the media, has involved such terms as *genetic inheritance* and *heritability* as if they were just two ways of saying the same thing. "Genetic inheritance" seems rather straightforward. It can apply to an individual, as in "he has his mother's eyes" or, perhaps still more loosely, "she has her uncle Clyde's temper." Obviously, statements of this sort can apply at a collective level as well, as in "that branch of the family has always been a little wacky." Needless to say, such statements may not always be accurate in a scientific sense.

The term *heritability* can also be used in a loose metaphorical sense, but as a scientific term it has a more precise meaning. For one thing, it applies at a collective, not an individual level. Just as important, it applies to variance within a group or defined population with regard to a particular trait.

Heritability, in that sense, would not apply to just how much of Sarah's fiery temper is a result of her uncle's genes as opposed to how much of it could be due to other factors, such as having been picked on by her brother when she was small. Nor, for that matter, would it apply to the fact that Sarah has type A blood.

It could, however, apply to the frequency of type A blood in the student population of Sarah's high school, for example. As we recall, heritability deals with variance within a population of some sort and refers to differences in the frequency or distribution of a particular trait. Thus, in the unlikely event that every student in Sarah's high school had type A blood, type A's heritability in that population would be zero because the variance was zero.

The basic information we would need to determine the degree to which differences in intelligence are genetically inherited remains either unavailable or heavily contested, even if we could agree on a definition of just what "intelligence" is. Heritability, even if it could appropriately apply to "intelligence," has to do with individual variation *within* groups, not differences in collective or average traits *between* groups.

We can approach the issue of the fixity of intelligence from another angle, however. As we've noted, typically, students in well-off suburban schools score higher collectively than students in poorer urban schools. It turns out that in cases where such suburban schools have received additional resources, academic metrics such as test scores haven't changed much. But when urban schools have benefitted from the same sorts of measures, scores have increased substantially.

It would seem, then, that whatever fixed abilities might have existed in both categories of students, the urban students' innate abilities were, in fact, far beyond what they had been able to achieve in a relatively impoverished school environment. On the other hand, students in richer environments had already been able to approach their potentials more closely, since they'd been working in favorable conditions to begin with. As to the question of whether social enhancement programs are a waste of money or actually make a difference, the conclusion seems obvious.[7]

LET'S GET THAT SOCIAL RANKING STRAIGHT

By the 1960s, this stuff was getting pretty old. The list of just who was innately "inferior" has shifted from the early years of this country. At various times it had been Irish, Africans, Germans, lower-class English, Slavs, Jews, Chinese, Italians to, most recently, immigrants from Mexico and Central America. Aside from that rich variety of humanity, the theme has been pretty constant. It's gone something like this: It's not our fault that these people have inborn tendencies toward depravity, drunkenness, promiscuity, larceny, disorderly conduct, or just aren't very smart. It isn't even their fault. It's just a result of their inherent "nature." And that being the case, there isn't much to be done about it except to keep them under control and/or to restrict their numbers.

As we know, measures to restrict population numbers at times have involved eugenics, including sterilization, clampdowns on immigration extending to such measures as building walls along national borders, and expulsion or deportation. (In the 2012 presidential election campaign, we also heard one candidate suggest a policy to encourage "self-deportation," apparently by creating an environment so unpleasant for immigrants that they'd choose to go back where they came from.)

In 2013, a bill to improve immigration policies stalled in Congress. Republicans insisted on expanding a wall along the border, doubling the number of enforcement officers, and purchasing assault helicopters and drones at a cost of about $30 billion. They insisted on these policies even though, in 2013, migration from Mexico had trickled to a net rate of zero.

By the summer of 2014, tens of thousands of children from Central America, many of them fleeing gang violence, rape, or the threat of murder, made their way to the southern border of the United States. In the face of what most people recognized as a humanitarian crisis, some politicians raised the alarm that they might be carrying disease or that the children posed a "threat to national security." One politician suggested that most of them were carrying illegal drugs.

There was a sense in some quarters that the number of immigrant children was overwhelming the country, even though by the end of August, their numbers amounted to the seating capacity of one good-sized football stadium. Many politicians insisted that they be sent back immediately—even though for many, this would result in almost certain death. One Congressman referred to this measure as "reuniting them with their parents"—even though in many cases, it had been their parents who had urged them to flee for their lives.

NOTHING IF NOT PERSISTENT

When the concept of IQ had come along, it had simply added another element to the mix. Using IQ made it possible to gauge the inferior nature of various populations with a metric that smacked of science. Numbers don't lie, even if people might use them to. But after all of this, despite the somber claims of so many researchers, no one has ever been able to establish firmly that any substantial amount of intelligence is biologically inherited, and therefore "fixed."

This does not, of course, apply to those unfortunate individuals who suffer some disability that impedes the normal functioning of the brain. And as anyone who has lived among other people for any length of time knows, not everyone is equally smart, no matter how we might care to define the term. But no research has been able to determine the precise role of biological inheritance in the functioning of normal intellect. This has not stopped IQ enthusiasts from going forward as if it had, however.

Many of the arguments from the 1970s onward have used the "if . . . , then . . ." framework. Harvard psychologist Richard Herrnstein made such a case in positing that a social meritocracy exists based on differential inborn natural abilities.[8] He argued in effect that *if* there is a substantial biological component to intelligence, and *if* higher intelligence is linked to success, *then* social classes reflect the natural order of things.

But "if," of course, is an admission of uncertainty. We might substitute a parallel, nonbiological set of propositions. *If* wealth is differentially distributed in society, and *if* wealth is inherited, and *if* wealth generally leads to better education and other social, economic, and political advantages, *then* the social order basically reflects the distribution of wealth.

We could take it further. Some proponents of the biological argument would elaborate: *if* intelligence is inherited, and *if* intelligent people tend to marry other intelligent people (or perhaps, we should say "mate with," since we're talking biology here), *then* their offspring, having genetically inherited good intelligence from both parents (it's hard to resist the temptation to use the term *intelligenes*) are also likely to excel in society. On the other hand, *if* wealth is inherited, and *if* wealthy people tend to marry other wealthy people, *then* their offspring, having the benefit of wealth from both parents, are likely to enjoy economic and social advantages. The reader may decide which of these is more persuasive, based on experience and observation.

Herrnstein later collaborated with Charles Murray, a prolific political scientist at the conservative Heritage Foundation, in writing a ponderous tome titled *The Bell Curve*.[9] The title refers to a normal distribution of data on a statistical graph, with high frequencies of whatever measurement is expressed in the center and lower frequencies at each end, producing an image that to some might resemble a sheet draped over an octopus.

In the pages of this tome, among a rain forest of verbal foliage, the same familiar assertions emerge. Since African Americans on average don't score as well on IQ tests as "whites," the authors assert, the most appropriate social policies should not involve attempts to change that. Indeed, the authors even suggest that such programs as Head Start could give rise to social unrest because of "white anger." Instead, they argue, the more promising strategy would be to help African Americans and other populations who score poorly on IQ tests to find productive endeavors more appropriate to their abilities. It's largely a matter of finding the right "fit."

By now, the pattern should be all too clear. We've seen the determination on the part of many of these writers to establish a case for the biological inheritance of intelligence whether the evidence supports it or not. It resembles an irrational article of faith in the face of insufficient data, a conviction that the evidence *must* be there, if only we could find it. In the meantime, we'll carry on as if we had.

This fondness for biological explanation goes far beyond the researchers who have pursued the enterprise directly. The news media have also shown a hearty appetite for such "findings." On several occasions, the *New York Times* has all but jumped up and down clapping its hands at some alleged breakthrough in that area.

In 1990, the headline of a *New York Times* story suggested that studies of twins now showed that IQ may be far more determined by inherited genes "than previously thought." The science editor did mention in passing that many researchers in the field questioned the findings. But never mind. This was an exciting development.[10] And we might recall that when researchers challenged Boas's earlier findings on head shape in 2002, which had found that the head shapes of children differed from those of their own immigrant parents, *Times* reporter Nicholas Wade covered the story with palpable excitement.[11] A rebuttal by other researchers in the *American Anthropologist* apparently didn't strike the press as being that interesting. As a consequence, many casual readers, who rarely, if ever, peruse the *American Anthropologist* continued to believe that Boas's findings were disproved.[12]

All of this raises the question of why the possibility of biological influence on intelligence has such popular appeal. Here we can only speculate.

When it comes to efforts to justify exclusionary policies, such as eugenics, deportation, or border fences, the tone is reminiscent of many other cultures' beliefs in mystical pollution.[13] Targeted groups have often had a dangerous aura to many—not necessarily threatening in a physical way, but "unclean" at a more abstract level—and liable to sully the inherent qualities of the predominant population through admixture, or miscegenation, or just by being around. (We might recall that "interracial" marriage has only become legal in all states within the last fifty years.)

These elements have tended to be more subtle in some of the later IQ debates. True, there is little apparent fondness for the low-scoring categories of people. But some authors, at least, have expressed concern for their well-being, offering crocodile tears over the challenges of their inadequacies. Perhaps some of this is strategic, a device for warding off accusations of racism or bigotry. But at heart, these assertions continue to be arguments for separation. More generally, one suspects, there's some comfort in the idea that what *is* is what *has* to be. No need to have a guilty conscience about something that's beyond our control, a result of natural forces greater than all of us.

TWINS, AGAIN!

Despite the discrediting of Sir Cyril Burt's conclusions, some researchers continue to push on down this particular road, no more daunted in their quest for biological explanation than the knights of old in their search for the Holy Grail. Foremost among them is Thomas J. Bouchard Jr. of the University of Minnesota. Bouchard amassed all of the cases he could find of twins who had been separated early in life. The database included the remarkable Ohio brothers with the blue Chevrolets, the dogs named Toy, and wives named Linda (and later, Betty).

In 1990, Bouchard and his colleagues published some of the results of their work.[14] They claimed to have demonstrated a strong genetic basis not only for intelligence but also for numerous other personal characteristics, ranging from emotional temperament to political attitudes. Some like-minded colleagues found Bouchard's work a basis for a final, definitive answer to the IQ/inheritance question. The media were titillated. Even the relatively serious and generally liberal *New York Review of Books* published a favorable article.[15] Not everyone was equally enthusiastic or convinced, however.[16]

For one thing, the manifestation of traits that have a genetic basis is not usually straightforward, except for genetically simple traits that depend on the presence or absence of a single gene. The ABO blood groups are probably the most common example of this.

Without going too deeply into the "birds and bees" issue, each person, as we know, gets one allele or version of a gene from each parent. This is the person's genotype. The type of blood one actually *has* is the phenotype. A genotype consisting of two alleles for A (AA) produces a phenotype of type A blood. A genotype of alleles for A and O (AO) produces type A blood because A is dominant and O is recessive. To have a phenotype of type O blood, one would need to have a genotype of two alleles for O (OO) or, in other words, to be homozygous for O. Since neither A nor B is dominant over the other, a person with an allele for each, a heterozygous genotype of AB, would have type AB blood as a phenotype. For a person to have type B blood, he or she would have to have a genotype that is homozygous for B (BB) or heterozygous with alleles for B and O.

But very few visible, let alone behavioral, characteristics are genetically simple. Most, if they have any significant genetic basis at all, depend on the interaction of multiple alleles. Moreover, genes and their interactions also involve environmental factors, with variable results. Nutrition, variations in temperature at critical times of development, chemical interactions, and a range of other contextual issues affect the ways in which genes interact and express themselves. And this is the case even for straightforward physical issues such as the sequential development of sexual characteristics, stature, obesity, and the onset of certain diseases with hereditary components. These are complex enough, and they continue to be a focus of research. How much more complex and subject to environmental interaction, then, are such factors as temperament, moral conscience, artistic ability, and, yes, all of the things subsumed under the rubric of "intelligence"?

As we've seen, many researchers, perhaps in despair of defining such an immensely complex phenomenon, have fallen back on IQ scores to provide a simple quantitative measure that's precise and therefore appears scientific. The argument has been, further, that such scores are relatively stable, both at the individual and group levels. But several studies have shown that even this simple measurement has been subject to change as a result of altering the environment—even in the case of children who are well past infancy and have already learned to read and write. In one case, IQ scores rose by about 20 percent.[17]

No reasonable person could argue against the existence of any genetic component at all in any human behavior, including "intelligence," however one might define it. But it also seems clear that with regard to such complex characteristics, genes provide the *capacity* for development in various directions rather than, as in the case of blood types, *determining* traits.

As for those pesky identical twins, it turns out that their genes don't necessarily dictate their destinies. A recent study of numerous pairs of identical twins raises important challenges to earlier studies and conclusions.[18] The researchers "found that young twins had almost identical epigenetic profiles but that with age their profiles became more and more divergent . . . the epigenetic profiles of twins who had been raised apart or had especially different life experiences—including nutritional habits, history of illness, physical activity, and use of tobacco, alcohol and drugs—differed more than those who had lived together longer or shared similar environments and experiences."[19]

HOT AIR FROM CANADA AND FROM ACROSS THE POND

Before leaving the feverish realm of IQ inheritance debates—although they remain with us, as we'll see in later chapters—

we should give credit to two other participants in the discussion. One of the more entertaining performers in the arena was the late J. Philippe Rushton, a psychologist at the University of Western Ontario.

Most of Rushton's material was the same old stuff: alleged "racial" differences in intelligence and so on. But Rushton introduced a slightly more ribald note to the issue. Unlike most other theorists in his camp, he placed a good deal of emphasis on the alleged sexual proclivities of the various "races."

Rushton was inspired by differences among nonhuman species with regard to their reproductive strategies. Some, like fish, produce large numbers of offspring but invest little energy in taking care them, relying on the likelihood that even though most of them won't survive, at least some of them will. This is sometimes referred to as "r selection." Others, such as most mammals, have fewer offspring but care for them more intensively. This is known as "K selection."

Although these terms originated with reference to the reproductive characteristics of species as different as fish and mammals, Rushton argued that these alternative strategies provide models that could apply within a single species, comparing various human populations. Africans, he suggested, were more like fish, producing lots of offspring but investing relatively little energy in caring for them. Europeans, on the other hand, had fewer offspring but invested more energy in caring for them. These contrasting reproductive strategies resulted in differing levels of IQ. And lest anyone mistake this for racism, Rushton, of European descent himself, placed Asians further still along the continuum of low birthrates, high parental investment, and high IQ levels.[20]

The latter point might seem a rather odd conclusion, considering the figures on population increase in Asia, although that, admittedly, may be a minor point. It pales in comparison with his apparent ignorance of child-rearing patterns in most cultures, and of those among the hundreds of societies of Africa in particular. But Rushton had still more to say about reproduction. As an aspect of Africans' alleged reproductive strategy, he asserted, their anatomy reflects this championship level of sexual activity. Rushton expounded at some length on the features of penis and buttock size. Basically, he argued, the r reproductive strategy led to evolutionary adaptations that included larger penises and smaller brains.

Rushton pretty much cornered the market on the topic of penis size and buttocks in the scientific literature on IQ inheritance. But wait— haven't we heard something like this before? Oh, yes. Remember Herbert Spencer, the nineteenth-century social philosopher who gave us the gift of social Darwinism? Spencer's model of progress through competition, we might recall, extended to the various cells within the body. The rate of reproduction among the upper classes was lower, he explained, because in the competition between brain cells and reproductive cells, the brain cells prevailed. Among the lower classes on the other hand—well, you remember the rest.

As we've seen so far, bad ideas can show remarkable persistence, even after lying relatively dormant for a century or more.

Another figure deserves mention. Hans Eysenck, a British psychologist, also became a major voice in the IQ discussions. Eysenck's basic argument was essentially the same: IQ is inherited; it differs among "racial" groups; social programs attempting to improve the scores of disadvantaged populations are foolishly misguided.[21]

We might marvel at the dogged persistence of these advocates of inherited IQ in the face of consistent refutations of their work within the scientific community. One factor, as we've seen, has been the generally favorable attention they've received in the media. The typical news coverage of various claims has used such terms as *breakthrough*, while acknowledgment of reservations among the established scientific community often conveys the image of stodgy professors guarding their turf, unwilling to accept brash new ideas. As we've already seen, however, these ideas may be brash, but they're hardly new.

We might also perceive some reinforcing camaraderie among this staunch band of brothers (with an occasional sister thrown in). Jensen was a student of Eysenck. Eysenck praised Jensen's work.[22] Jensen and Rushton coauthored a paper in 2005.[23]

On occasion, they suffered unfair and sometimes even violent assaults in public venues because of their work. Unfortunately, these unconscionable actions on the part of those who vehemently objected to what they had to say only gave them the status of heroic martyrs in their own eyes and the eyes of their followers. In the public arena, such attacks also served to obscure the many valid scholarly objections to their conclusions.

We should also note a major source of financial backing they received. Those of us who enjoy conspiracy theories can be gratified to find that there was, indeed, mutual reinforcement here among like-minded people, and considerable financial support for them. In this collective effort, the Pioneer Fund has been a major player.

EVERYONE NEEDS A FRIEND

Begun in 1937 as a response to concerns about immigration, the Pioneer Fund at that time evinced a strong interest in eugenics. According to the Southern Poverty Law Center, the fund's "original mandate was to pursue 'race betterment' by promoting the genetic stock of those 'deemed to be descended predominantly from white persons who settled in the original thirteen states prior to the adoption of the Constitution.' "[24] Some of the fund's members expressed sympathies with Nazi policies of the era, but following World War II, the Pioneer Fund officially denied any Nazi connections. From the second half of the twentieth century to the present, however, the Pioneer Fund has supported the work of numerous researchers engaged in issues of "heredity and human differences."[25]

The philosophy of the Pioneer Fund appears in the work it has supported. Many of the recipients of Pioneer Fund grants sound like old acquaintances by now: William Shockley, Arthur Jensen, Hans Eysenck, J. Philippe Rushton, and Thomas J. Bouchard Jr. of the Minnesota Twins study. In 2002, Rushton took over the helm as president of the fund.

Although attributing "guilt by association" is notoriously unjust, we could nonetheless refer to another old saw: One can be judged by the company one keeps (and supports financially). In this case, however, a far better basis for judgment has been the work, the assertions, and the conclusions of these scholars. As we've noted, their basic paradigm— that intelligence is biologically determined, that it differs among "racial" groups, and that it's unlikely to change through social programs or other environmental factors—has suffered not only criticism but rebuttal and downright refutation on a number of scores.

Perhaps it's time to leave this dreary battlefield and move on to something with a different focus. Sociobiology in the 1970s continued the theme of strong biological influence on behavior, but we can feel some relief that at least "race" no longer held center stage.

Chapter 5

SOCIOBIOLOGY: A NEW SCIENCE OF THE SAME OLD THING

MUCH OF THE DISCUSSION ABOUT IQ INVOLVED REFER-
ence to the processes of human evolution. Purported
genetic differences among "races" rested on the assumption that whatever evolutionary experiences had differentiated regional populations
deep in the past had led to differences in IQ scores in the present. And
these differences, being the results of long processes of natural selection, could hardly respond to short-term social measures such as
remedial educational programs. Although some proponents of "racial"
IQ differences didn't spend much time addressing this aspect of their
position, their conclusions inevitably rested on such assumptions.

J. Philippe Rushton, of course, was far more explicit than most. He
argued that as early humans migrated out of Africa, those who eventually peopled Europe faced greater challenges for survival than their former relatives in tropical climes; therefore, they had to use their wits to
survive. No problem seeing where that was leading.

Little of this rose above the level of fable. True, early humans
migrated from Africa to Europe and other parts of the world. And yes,
Europe in general is colder than Africa. But the rest is essentially a "just
so" story, not much different from folk tales about how the leopard got
its spots or why bunnies have short fuzzy tails. In fact, there probably
are some sound evolutionary theories to account for those phenomena,
based on evidence. But we have no evidence whatsoever about the relative intelligence of various prehistoric human populations, aside from
the overall increasing complexity of tools and other artifacts through
time and the appearance of graphic art in the Paleolithic.

By the 1970s, the "race"/IQ issue, persistent as it remained in some quarters, had become tiresome and downright distasteful to many. Yet a much broader base of interest in biological influences on human behavior persisted. That interest, for a time, became the basis for the new field of sociobiology.

"GO TO THE ANT, THOU SLUGGARD. CONSIDER HER WAYS AND BE WISE"

The founder of sociobiology was a Harvard biologist named E. O. Wilson. Wilson had a distinguished reputation as one of the world's leading authorities on ants and was a renowned expert on bees. Both of these, of course, are social insects. They survive through an intricate coordination of activities and collective responsibilities that allow the colony, or community, if we can call it that, to exist. And as far as we can tell, these activities arise on the basis of genetic imperatives rather than on what we humans call "learning." As the biologist S. H. Waddington put it, "Wilson is perhaps the leading authority in the world on the social life of insects, and insects are notoriously unteachable creatures."[1]

Subspecies of these insects may differ in their patterns, but the consensus view is that these are genetic variants, not what one might call ant or bee "cultures" that could change over a few generations as human cultures do.

In the 1970s, Wilson began to apply what he had learned from ants and bees to a broader arena. As we know, humans, like these insects, are also inherently social creatures. Even the occasional hermit *is* a hermit because unlike almost everyone else, a hermit lives outside of society. To be a hermit, one needs a society to exist in the first place, so one can then reject it. A solitary existence is possible, of course, but it's generally an aberration or a punishment, not the human norm. Even someone who's been banished from one society is liable to end up in another one, if he or she survives at all.

Like populations of ants and bees, human society operates on the basis of coordinated, complementary activities. Sure, humans also have conflicts. But so do ants. And humans, like ants and bees, also have genes. So what can one species tell us about the other?

Wilson's tome *Sociobiology: The New Synthesis*[2] was an impressive work, filled with ample documentation. Most of it dealt with nonhuman species, and up to chapter 27—the first chapter to deal with humans—it provoked little controversy. Much of the work was a continuation of Lorenz's work in ethology, the science of animal behavior, with more than a dollop of genetics thrown in. Other ethologists and biologists might have had some quibbles here and there, but it was mainly when Wilson ventured into the subject of *Homo sapiens* that cries of outrage arose.

Equipped with the knowledge that the complex workings of ant life rest on genetic codes, Wilson set out to examine the panorama of human behavior in search of things that might also be driven by genes. It didn't take him long to find some. As everyone knows, people have different personalities and differing abilities in many things. Wilson became convinced that many individual traits—shyness or aggressive behavior, for example—are due to the presence or absence of particular genes.

The root causes of many of these complex human traits remain mysterious to most of us. Even those of us who would emphasize the importance of experience and learning are likely to recognize that people differ at what is probably an innate level with regard to musical or artistic abilities (what we often refer to as "talent") and in many other respects. Certainly, most personal characteristics involve interactions between the genetic "raw material" in a person's genotype and the experience, whether traumatic or enhancing, a person undergoes through the course of a lifetime and what she or he learns from it—consciously or unconsciously. One question, for the most part unresolved, is the degree to which these factors interact to produce the results we can observe in a living person.

Wilson, being a biologist with an interest in the genetic bases for animal behavior, looked for biological causes in the realm of human behavior as well. This is not to say that he found them; but in many cases, he did assume them and claim their existence. Some of these claims were more controversial than others. The issue of "male domi-nance," for example, caused considerable discussion at the time.

EDWARD, HAVE YOU MET HERBERT?

In many respects Wilson's perspective is consistent with the views of the nineteenth-century sociologist Herbert Spencer, whose ideas we've encountered earlier. Spencer saw society as a gigantic organism, and he often receives credit for developing what we refer to as the "organic analogy." In Spencer's view, however, it was more than an analogy. Spencer went beyond claiming that society was *like* an organism. His arguments slid smoothly into explaining why society *is* a sort of organism, albeit of a special type. Indeed, he went so far as to compare small societies to simple, one-celled organisms (think of amoebas) and to compare complex state societies to complex multicelled organisms such as mammals.

Spencer had no problem coming up with vivid examples to illustrate his points. In his view of the world, telegraph lines functioned as nerves, railways and other transportation routes provided circulation, and so on. Not only that, but societies could grow, compete with one another, and even die. And the key dynamic process that kept things humming was competition at every level. Competition not only assured the survival of the fittest among organisms and species in nature. It applied to societies as well.

To modern readers this might seem to be little more than creative imagery. There is, of course, no empirical evidence to support most of Spencer's depictions of society as a giant organism, and public opinions on the matter have pretty much moved beyond that—largely in favor of an image of untrammeled individualism (which may be just as off-target in the other direction).

To Spencer, however, this was more than a metaphor. It was an analysis of reality. The analogy (or analysis) was so reasonable and convincing, he apparently believed, that to challenge it would be quite *un*reasonable. In that vein, rather than seeing competition within and among societies as analogous to dynamics occurring in nature, Spencer considered societies themselves to be *part* of nature, operating according to the same natural laws.

This had many implications for the proper study of social phenomena. Traditionally, attempts to understand the human experience had involved the study of history, the interplay of ideas through time, politics, economic forces and policies, and so on. The study of humans generally had involved the study of what humans as individuals or collectivities have thought, done, created, and destroyed, how they've interacted, and for what motives. In Spencer's view, though, the important question was how humans responded collectively and individually to the laws of nature.

Spencer believed that the more traditional academic approaches to understanding the human experience were on the verge of becoming obsolete in the face of new understandings and Darwinian insights. All humans and other species, at the most fundamental level, follow the principles of biological evolution: principally, selective pressures and competition for survival. Eventually, Spencer argued, the other disciplines, whether they focused on society, history, the humanities, or even the arts, would eventually become subsumed within a single biologically informed body of theory.

Wilson did not acknowledge any strong inspiration from Spencer's ideas. In 1975, a century or so after the height of Spencer's influence, this would probably have come across more as a confession rather than as a point of pride. But the two of them would almost certainly have had a lot to talk about. On the matter of restructuring academic disciplines alone, for example, Wilson almost seemed to echo Spencer in asserting that "sociology and social sciences, as well as the humanities . . . are branches of biology."[3]

QUIET . . . I THINK THE GENES ARE UP TO SOMETHING

If behavioral patterns express genetic signals, messages, or directions, the genes producing these patterns must have evolved over a long period of time, over many hundreds of generations, as a result of natural selection. Somehow, individuals possessing those particular genes must have been more successful than others with alternative genes in the struggle to reproduce viable, fertile offspring, thereby passing their own genes onto future generations. Genes that have come down to us in the present must have been the ones that prevailed in competition with other genes.

If this sounds a bit like anthropomorphizing genes by attributing some sort of intentionality to clumps of molecules, the argument did sometimes approach that form. The sociobiologist Richard Dawkins wrote a book titled *The Selfish Gene.*[4] From some sociobiologists, we get the idea that the genes, rather than the organisms that carry them, are in fact the major players (or competitors) in the arena.

It would seem that the organisms that carry the genes, the complex beings that live, search for food, reproduce, and die, are basically little more than vehicles that transmit their genes from one generation to the next. Such a model, of course, leaves little room for the operation of many phenomena that appear to be unique to the human species—language, culture, and the ability to process and create information with high levels of complexity. To quote C. H. Waddington once again referring to Wilson's *Sociobiology*, "Is it not surprising that in a book of 700 large pages about social behavior there is no explicit mention whatever of mentality?"[5]

This model of competitive, selfish genes involved some inherent logical problems. In real life, individuals sometimes act in a way that's likely to be detrimental to their own survival in order to improve the survival chances of others. In the human realm, we sometimes refer to this as heroism. Such behavior might even become the stuff of legend. Comparable behavior occurs in other species as well. But is it really the same thing?

As a classic example, suppose a cat sneaks up on a flock of birds. One of the birds might spot it before the others do and raise the alarm, squawking and fluttering and allowing the rest to escape. It's hard to see this as some sort of ornithological valor, though, since the squawking bird is most likely frightened out of its wits and just doing what birds generally do when they feel that way. Most likely, it just wants to get out of there. Having been the first to see the cat and fly away with all of its genes intact, that fortunate bird might be the one most likely to contribute to the next nest-full of gaping little beaks. It might be a little different if the bird actually went after the cat, gamely holding it off till the others had time to escape.

Still, some sociobiologists have suggested that the bird is placing itself in greater danger than the rest of the flock by drawing attention to itself rather than, perhaps, silently tiptoeing away. Such risky behavior, they suggest, exposing itself to the cat, makes it less likely that this flustered bird will be able to pass on its own genes. Given the assumption that the bird's behavior is genetically driven, are the genes of that bird somehow less selfish?

Perhaps a better example would be an animal that actually confronts a predator, thereby, unambiguously placing itself in greater danger than the rest of the group that it's trying to defend.

Whatever the example, instances of animal behavior that resemble human altruism pose a problem. If the organism is basically little more than a vehicle to pass on its genes to the next generation, how can we make sense of such cases of apparent self-sacrifice?

Not to worry. Sociobiologists came up with the idea of "inclusive fitness." A bird or other animal might place itself in danger to protect the rest of the group because normally, they would all be close relatives and share many of the same genes. The sneaky cat might end up with feathers dangling from its whiskers, but most of that unfortunate bird's genes would have flown off to a safer place. This principle became a centerpiece of sociobiological argument by turning a potential problem into supporting "evidence" for the underlying premise of gene perpetuation at all costs. Nicely done.

There are problems in conflating the reflexive behavior of birds and other nonhuman species with human altruism, of course. For one thing, human altruism involves intentionality and complex decision making— even if one might later describe it as a "split-second" decision. Such behavior also varies greatly among individuals and might even vary for the same individual on different occasions. Just about any bird is likely to fly away squawking when it sees a cat. But not everybody is likely to be a hero. That's one reason we often give special recognition to those who are.

WHAT'S GOOD FOR THE GOOSE

Like Lorenz and others, Wilson proposed a continuum between humans and other species. So far, so good. It would be difficult, perhaps even absurd, to argue that we're not at some level just another animal species (setting theological discussions aside, of course). On the other hand, lots of species have developed specialized abilities. Bats, unlike other mammals, can fly around in the dark, detect flying insects through echolocation, catch them, and eat them. Sharks can sense electromagnetic impulses from fish hundreds of yards away. Humans have developed a large degree of flexibility in their behavior through a heightened learning ability and the use of language. This allows us to retain and accumulate information and then transmit what we've learned to others through the use of symbols—whether voiced, gestured, or expressed through abstract graphic squiggles of various sorts. No other species has this specialized, biologically evolved capacity to the same degree. (We probably should give a nod here to whales and dolphins, which seem to have some fairly complex vocal communication going on.)

We might choose to be proud of that fact or merely take it for granted. Wilson and other sociobiologists largely chose to ignore it. For them, the most salient aspect—and cause—of human behavior lay in the genes. At one point, Wilson even suggested a genetic basis for a propensity to learn one language rather than another. Along the way, he and others posited many genetic bases for other rather specific aspects of behavior as genetically driven. By now, the themes of aggression and male dominance are all too familiar. It seems unfortunate that we need to continue revisiting these old tropes. We'll stop when they stop.

At this point we could probably draw back and take a deep breath or two. Certainly, the forms of behavior that sociobiologists discuss *are* part of the human repertoire. That fact in itself leaves no doubt that humans have a genetic capacity to act that way. Let's, for the sake of discussion, consider "aggression" to include anything from threatening behavior to physical attacks. No one needs to be told that humans can and often do act aggressively and that we have genes that allow us to do so. The more important question has to do with what factors *produce* such behavior.

We know that some chemicals can have that effect, so we can't entirely rule out a physical basis. We also know that myriad other causes can give rise to aggressive behavior and that individuals differ in the degree to which they manifest aggression. All of us probably know someone who "wouldn't hurt a fly" and others who "have a hair trigger." We've also heard of individuals who rarely display aggression but suddenly explode in rage for some reason: "He was a quiet boy . . ."

Populations with different cultures also vary in this regard. Some value displays of bravado and self-assertion. Others prize self-control and the ability to maintain equanimity in the face of stress. It seems evident that, in either context, a major aspect of a child's rearing is to instill these ideal forms of behavior, whether they become an inherent part of the individual's personality or become superimpositions on a more basic personality, or some combination of both. It also is clear that factors producing these observable phenomena are many and complex.

These examples, and many others we might consider, certainly are manifestations of genetic potentials, physical and behavioral capabilities that ultimately have their roots in the genes. But how much does that tell us? How useful is it for sociobiologists to claim that humans have a genetically based potential for aggression? One might as well argue that since most humans inherit two functioning legs, they have a genetic propensity to walk. But walk where? With whom? For what purpose? In sneakers, high heels, or skis? Where does dancing come in?

Even more questionably, sociobiologists have suggested that human populations differ in the degree to which they have a genetic propensity for violence. This reliance on genetics as a primary cause would, of course, subsume such issues as history, politics, and ideology, all of which involve changing, unstable, and interacting complex factors. Unless, of course, such historic episodes as Alexander the Great's conquest of a large part of the world, the Renaissance, or the Industrial Revolution were genetically driven. Really?

While we're at it, let's take another look at the idea of "male dominance." Here we do have some basic biological factors to consider. Adult males in general are larger than women, and usually they're physically stronger, at least in terms of short bursts of energy. Moreover, the biology of reproduction means that women are liable either to be pregnant or nursing infants during long periods of their most physically active and productive years, while males' primary role in the biological process—assisting in conception—is far less complicated. How much can we glean from that?

There's no doubt that the powers of men and women in many societies are unequal, partly, perhaps, because of those "facts of life" we've just enumerated. But not always. In many societies, women hold considerable power. Some societies are essentially egalitarian.[6] Typically, men and women hold and exercise authority and responsibility over different aspects of life—authority that might be compartmentalized, but nonetheless complementary and equivalent. Yet in all of these societies, sexual dimorphism and the biological processes of reproduction are the same. The point is that social, economic, and cultural circumstances act in different ways on a common genetic basis—a basis that allows, and in fact requires, considerable flexibility.

WAIT . . . ARE WE STILL DOING SCIENCE?

And here we come to a basic question. Many prominent sociobiologists such as Wilson have built distinguished careers as people of science. Science, at a very basic level, depends on empirical evidence. It may involve careful observations of naturally occurring phenomena or controlled, replicable laboratory tests and experiments. Even field studies occur under conditions that require as much accuracy and verification as possible. Why, then, when these distinguished scientists begin discussing the general human experience, does the door swing wide open to speculation?

At some point, as we noted many pages ago, the enterprise shifts from science to social philosophy. This is all well and good, up to a point. Social philosophy is a favorite pastime of humans everywhere.

But unfortunately, these philosophers seem to have neglected to take the "science" ID tag off the windshield. They continue to occupy "science" parking spaces, but they seem to feel exempt from the rules.

One problem is that, despite their frequent assertions that particular forms of behavior such as aggression or shyness have a genetic basis, no genetic locus for a particular form of behavior has ever been identified. For one thing, most behavior has complex causes. It might be one thing to consider a simple reflex, such as sneezing, to have a genetic basis. We don't have to learn how to sneeze; only how to do it as politely as possible. We might even learn, depending on our culture, that it's good luck, or bad luck, or that there's something you should say when you do it or when someone else does it. But that sort of digresses into cultural stuff, doesn't it?

Other than reflexive acts like sneezing or coughing, few of us ever do anything for a single reason. Anything we do is a result, at one level, of a genetic capacity; otherwise, we wouldn't be able to do it at all. But knowing that we have inherited the capacity to speak tells us nothing about what we're liable to say or even in what language we're likely to say it.

Given that what we do or how we act begins with our capacity to do so, what else might be involved? What makes us do what we do? The way we've been taught we're supposed to act in given a situation? The sort of mood we're in that day because of what we've had or haven't had to eat? How well we slept the night before? What somebody said to us an hour ago? What we've seen someone do in a movie? The impression we want to give someone nearby? The list of possible factors can be almost endless. Most of them have nothing to do with genes except at the most basic level.

Wilson attempts to fend off some of this obvious criticism and still keep genes in the picture. He asserts, for example, that there are "genes promoting flexibility in social behavior at the individual level."[7] Nice try. But why would we have to invent "genes promoting flexibility"? Why not an absence of genes that compel specific forms of behavior in the first place?

At other points, he takes a harder position. As we've discussed, most modern scholars see cultural differences among human populations as clear examples of the flexibility of learned behavior. For the last century or so, this has rested on a premise of the biological equivalence of all human populations. Since the biological realm is a constant, the variation in custom and behavior obviously results from other factors. To take an old example, a child born in China who comes to the United States at an early age and grows up in a typical American family will grow up acting like a typical American kid, not like a typical Chinese kid. Nor, as numerous documented cases of rapid culture change demonstrate, are culture patterns "fixed" in any biological way. Modern Japan is quite different now from the way it was in the 1930s, let alone in the days of the shoguns.

But Wilson, apparently, isn't so sure. "Variations in the rules among human cultures, however slight," he asserts, "might provide clues to underlying genetic differences."[8] Seriously? This idea has been discredited in anthropology and the other social sciences for several generations now, based on extensive, painstaking work with scores of cultural systems. But never mind. Wilson soon backtracks a bit. "What has evolved," he explains, "is the capacity for culture, indeed the overwhelming tendency to develop one culture or another."[9]

So that explains it. But wait; is this "need for culture" a matter of genetic programming or the result of a *lack* of genetic programming? Could it be that, unlike ants or bees, humans have developed a means of survival that involves the ability to learn different ways of interacting with the environment and with each other *without* the constraints of genes compelling specific kinds of behavior? No doubt genes are involved in this capacity to learn. But having done so, they've relinquished the steering wheel to environment and experience. With regard to biological success, this has worked pretty well so far. Humans have overrun the earth in the past few centuries, ranging far beyond their original tropical habitat to the polar regions, even spending time living underwater and in space. Whether this is good for the well-being of the planet is another matter. But let's see ants and bees do that, with their fancy-schmancy genetically determined behavior.

PULL UP A CHAIR; IT'S STORY TIME

Sociobiologists' means of telling their story uses the trappings of science, but it is, after all, just a story. And as they tell and retell the story, several narrative devices tend to recur. One is the use of metaphor. Genes "struggle" for survival, despite being molecular blobs that have no capacity for consciousness or volition. They're "selfish." They "compete" with one another. These are interesting ways of expressing interactions among mindless entities, but is there a danger, at times, that we might forget that these are only figures of speech?

Another form of anthropomorphism—the attribution of human behavior to nonhuman entities—involves referring to animal behavior in human terms.[10] One well-known study of aggressive mating behavior among bees, for example, refers to the process as "bee rape."[11]

We needn't waste too much time exploring the many things that are wrong about this usage, such as whether normal insect mating behavior, which actually is genetically driven, is analogous to a human crime. A more serious issue is whether criminal behavior in the human realm could be equated, even metaphorically, with the genetically driven behavior of an insect. Is a rapist just doing what comes naturally? If so, are most human males, who would never consider rape, deviant or genetically deficient in some way?

In one sense such questions are hardly worthy of much attention, given their absurdity. But they do reflect some of the decidedly unscientific aspects of much sociobiological writing, if science requires precision and accuracy. The common practice of applying terms that have meaning only in the human realm—war, incest, murder, nepotism, for example—and applying them to nonhuman species implies a continuum of sorts, a biological imperative underlying these distinctly human issues. And such a continuum is, in fact, precisely what Wilson and other sociobiologists assert.[12]

But the nonhuman animal kingdom isn't the only source of biological insight for them. Many also look to human prehistory as a rich source of just-so stories. This often involves a variant of the "if . . . , then . . ." logic discussed in the previous chapter. In this case, however, we could substitute a "since" for the "if" with regard to the initial, fundamental starting point of the premise. The reasoning is that "since" a given form of behavior *has* a genetic basis, *then* it must have originated earlier in human evolution.

As Wilson puts it, sociobiology amounts to "the systematic study of the biological basis of all social behavior."[13] Any questions? The fun part is making up what "must have" helped to bring this about, since unfortunately, our knowledge of the intricacies of life in the earlier stages of human evolution is rather scant. We saw an earlier version of this in the work of Robert Ardrey.

Although we don't know nearly as much as we'd like to about early human behavior, some writers have found an alternative source of insight: so-called primitive peoples, or, to be more tactful, "hunters and gatherers."

It's clear that agriculture is a fairly recent development in human history—perhaps in the last ten thousand years or so, give or take a few millennia. Since humans have been around much longer than that, it's evident that the bulk of human evolution took place in an era when all humans lived entirely on foods that were already present in their environment. They picked berries. They caught fish. They dug roots. They hunted. They gathered.

We don't really have much information about how these people deep in the past behaved. We've found traces of where they lived. We have some of their bones. We've found remains of a few of their ancient fires, sometimes with remnants of seedpods, fish scales, animal bones, clam shells, and other prehistoric trash that, because of their age and rarity, have become the stuff of archaeological treasure. Beyond this, we still don't have much to tell us exactly what these humans were like or how they behaved.

But wait. There have been people in more recent times who have lived by hunting and gathering. Why not use them as models for what these early humans were like?

THE GENERIC "PRIMITIVE"

There are a few problems with this. Almost without exception, hunting peoples in historic times have lived in areas unsuitable for agriculture or animal husbandry—too dry, too cold, or whatever. In some cases, they've ended up in such areas as refugees, pushed into them by expanding agricultural or herding peoples who, if not necessarily more aggressive, were certainly more numerous because of their food supply. Humans who lived before agriculture, though, had the entire habitable world at their disposal. Why would they have chosen to live in deserts when they had access to much more congenial, richer environments? Still, at least some of the strategies that have worked for historic hunting and gathering people might also have worked for ancient people making a living the same way.

To an extent, perhaps. But to look at a few well-documented cases of hunting peoples in more recent times—the San of southern Africa, for example, the Inuit of the central Canadian Arctic Coast, Australian peoples of the Western Desert, or Dene of the western Canadian subarctic—one thing that strikes us is their cultural diversity. True, they do have some things in common: They've had low population densities, and they've needed to be mobile in order to subsist on thinly scattered and often changing food sources in an environment of meager resources. This has placed constraints on what options were feasible for them.

But remember, hunters of the Pleistocene were not restricted to environments that nobody else wanted. Perhaps the best historic examples we have of nonagricultural peoples inhabiting a rich environment are on the North Pacific Coast of North America. There, such groups as the Haida, Tlingit, and Kwakwala speakers had relatively dense populations with large communal cedar plank buildings, elaborate systems of social ranking, distinctive graphic art, and complex ceremonial systems. All of this, and more, was possible because of their rich food supply of fish, especially salmon, and sea mammals.

The bottom line is that historic hunting peoples don't tell us much that we would like to know about life in the Pleistocene. These are not our living ancestors. They're not people locked into some sort of time warp, as some writers seem to have imagined. Their histories are every bit as long and as deep as anyone else's. And beyond some general similarities in organization, their cultures have been quite diverse. But that hasn't stopped sociobiologists from using these sources as rich fertilizer for the imagination.[14]

There's a certain irony to this. On the one hand, there's been a tendency to visualize a generic "primitive" way of life that blurs the distinction between historic nonindustrial societies and our prehistoric ancestors. On the other hand, as we'll discuss below, there's been little attempt in the course of this enterprise to pay much detailed attention to other living cultures when discussing an allegedly universal "human nature."

Facile references to our generic hunter-gatherer ancestors persist. In February 2013, the anthropologist Helen Fisher appeared on a television show called *The Cycle* on MSNBC. Fisher is somewhat unusual among anthropologists in embracing a sociobiological perspective. On that particular day, panelists were discussing a recent survey about dating preferences among American women. The results of the survey indicated that the three things women considered to be most important in a prospective date were a man's hair, his grammar skills, and his teeth.

The grammar part might be a bit puzzling—although perhaps some might find it difficult to resist a smooth talker. The hair and teeth, maybe not so much. One could imagine that bad hair and bad teeth could very well be a turnoff; no surprise there. The preference for good teeth, though, Fisher assured the panel, was a carryover from prehistoric times, when females saw this as a sign of good health.

Okay, maybe. It seems plausible. But is there really any concrete evidence for this? As a matter of fact, in early human history before the use of fire to cook meat, and thereby tenderize it,[15] bad teeth could not only mean bad luck with the ladies. It could also mean starvation, unless one could get somebody else to chew the meat up first. But let's not get into that.

AMERICANS AND BRITS: THE GENERIC HUMANS

Sociobiologists generally have paid little attention to cultural differences in behavior. There has been a tendency to see Western populations, particularly American or British, as examples of generic human beings.

Desmond Morris, whose "naked ape" work we discussed in chapter 2, carried this to an extreme, at least in terms of candor. When critics pointed out that many of the human gestures and mannerisms that he attributed to our inheritance from our primate ancestors are not characteristic of people in many other cultures, he brushed the objections aside. These other groups, he responded, were not really in the mainstream of human evolution. Aside from the elusive logic of this rebuttal, his view seems to reflect an attitude that has not been unique to him.

We might be inclined to excuse these omissions. Most sociobiologists, after all, are not anthropologists. It might not be surprising that they don't display a command of the myriad cultural differences that characterize human populations. On the other hand, there are lots of anthropologists out there, and one thing they like to do is to publish their work. Fairly detailed information on many peoples of the world is available to anyone with access to a good library. Sociobiologists are, after all, writing about the behavior of humans and advancing claims about its causes. Why not make an effort to find out more about the subject?

Implicit in sociobiologists' claims, however, is the assertion that cultural differences are relatively unimportant. Wilson uses the term "superficial" in referring to cultural differences and suggests that such diversity amounts to a relatively trivial veneer overlaying a deeper biological reality.

In one sense, cultural anthropologists might be happy to agree with this, if it means that biologically, humans are pretty much the same everywhere. That's really not Wilson's message, though. He holds that significant differences exist among humans because of differences in biology. While he doesn't explicitly link this to "race," it does have a familiar ring to it.

Wilson does make some reference to a few non-Western cultures. He alludes, for example, to hunting and gathering societies, such as the San people of southern Africa. But, he takes some pains to refute the conventional image of these people as essentially egalitarian—an image based on extensive fieldwork by numerous anthropologists.[16] He distorts and selectively cherry-picks the published information on them. At one point, arguing for a genetic propensity for people to sort themselves out as "winners and losers," he asserts that in this generally unstratified society where no one, until recently, had much property of any sort beyond items of clothing, hunting and digging tools, bags, and perhaps a few simple decorations, a few individuals tended to hoard wealth secretly without sharing it.[17]

There's no doubt that after the government of South Africa in recent years stifled San people's movements by restricting them in small reserves and interfering with their ability to hunt, their earlier social patterns broke down. As a formerly free-ranging, egalitarian society now trapped in a larger state system at the lowest echelon, the people were compelled for the first time to experience poverty as we know it, as a powerless minority in an industrial nation state with a cash economy. In that context, some of the earlier patterns of exchange and redistribution of meager resources broke down.[18] But this was hardly an example of pristine human nature, as some sociobiologists might have imagined it. More accurately, it was an example of a formerly successful way of life that had become incapable of functioning as it once had, as a result of oppressive governmental constraints.

Eventually the sociobiologists' heyday passed, partly in the face of a consistent chorus of criticism from other disciplines, including anthropology. By the late 1980s, the term *sociobiology* had all but disappeared from the literature except as a historic reference. As we'll see in the next chapter, however, its basic premises continued under a new name: *evolutionary psychology*. Its underlying assumptions remained part of the intellectual stew, continually appearing in the media.

A GENETIC GUIDE TO BEHAVIOR

One idea that had been an important aspect of the sociobiological scriptures was the assertion that women, for genetic reasons, are more conservative (selective? judicious?) than men in their sexual activities. The alleged reason, of course, had to do with passing on genes. Since a woman is unlikely to bear more than one child a year, when she allows herself to become pregnant she's making an important commitment to genetic posterity. Men, however, are freer to sow their wild oats hither and yon.[19]

From a sociobiological perspective, since women have fewer chances to pass on their genes than men do, they have to exercise a different sort of strategy. Essentially, so the story goes, a woman needs to choose a partner who not only has good genes but also is most likely to provide well for her and their children. This assumes, of course, that it's the men who provide for everyone else. Good genes aside, we should note that in many historical hunting and gathering peoples, which sociobiologists enjoy using as examples, it's women who contribute the bulk of the food supply and the most reliable portion of it.[20]

Women also, the argument goes, are looking for men with "good genes" to father their children. Actual women as persons may not always see things the sociobiology way, of course. Personality, career prospects, physical attraction, a nice car, or even a cool foreign accent could carry the day. Many a scoundrel has seduced a fair maid. But as any decent sociobiologist would quickly point out, many of these aspects of superficial attraction may in fact be indications of good genes, which is what it really is all about. This, as we'll discuss below, is one of those common assertions in sociobiology that can rarely be falsified and hence has no real scientific value. That which explains everything explains nothing. Even if that dashing hottie turns out to have lousy genes after all (whatever that may mean), one could still argue that the woman fell for him because something about him led her to conclude (subconsciously, perhaps) that his (invisible) genes were better than they actually were. Try and disprove that one!

Let's also recall that in many cases, people, both men and women, engage in sex with no wish whatsoever to bring about a pregnancy. In fact, they're liable to take whatever measures they can to prevent it. It's no accident that birth-control products are of great interest in American society. Moreover, in the most prosperous countries, where people are able to exert more control over their reproduction, birthrates have fallen, alarming some nativist politicians who fear being "overrun" by rapidly reproducing foreign immigrants. Such concerns, as we discussed earlier, have been part of the American experience throughout history. Of course, this could just mean that the people of Italy, France, and Spain, to cite a few examples, just aren't having much sex anymore. Maybe.

One would think that from a sociobiological perspective, the most powerful, wealthy, and privileged in society would pass on more of their genes than anyone else, because they can. Apparently not. Sociobiologists might adopt a "sure, now" position and argue that it would have been different during the Pleistocene. But if that good old Pleistocene strategy doesn't apply anymore, why are we even talking about it as if it did?

Part of the sociobiologists' scenario is also that men still have that Pleistocene tendency toward more profligate sexual behavior than women. But wait a minute. According to the *New York Times*, a recent study has revealed that, in the United States, at least, women are just as sexually active, with as many different partners, as men—and in some cases, even more so.[21]

Could it be that women just tend to be more discreet about it in response to cultural and historical issues? Anyone who has grown up in the United States over the past several decades can hardly be unaware of the "reputation" disparity associated with sexual activity, from at least middle school onward. Even in this supposedly more enlightened era, the ugly term *slut-shaming* has appeared in public discourse. It seems clear that if anything, this long-standing pattern would have led young women in any survey to underreport sexual activity to most investigators and young men—in at least some cases—to overreport. Researchers in this study found that, when they told women who had volunteered to participate that they were connected to a lie detector (they weren't), their reported sexual activity slightly exceeded that of men. A more recent article in the *New York Times* discussing sex without committed relationships (or "hookup culture") among academically high-achieving women students[22] provoked a number of responses, many of them questioning why this was even considered news.

Those who prefer seeing a genetic basis for behavior could dismiss this by pointing to the effects of our changing social mores and write it off as a recent development—an aberration, perhaps. But then we're left with positing a biological imperative that people often don't feel compelled to follow, which makes it somewhat less than imperative. Perhaps the only fallback position would be to assert that genes influence this aspect of sexual behavior . . . sort of . . . except when they don't—and that people can learn to behave differently from the way they used to only a generation or two ago. Did their genes change?

A comparable case arose recently with regard to a genetic locus that supposedly affects the way individuals react to stress. According to the *New York Times*, again, those who have two alleles for one version of a particular gene tend to become very stressed when confronted with a demanding situation, such as taking an important exam.[23] Those who have two alleles for the other version of the gene tend to welcome such challenges and rise to the occasion. According to the authors of the article, the researchers referred to these types of people, respectively, as "worriers" and "warriors." Supposedly, people with one allele for each genetic variant fall somewhere in between the two extremes.

But it turns out that it's not quite that simple. The researchers found that homozygous "worriers," those with two alleles for the "worry" gene, who received a pep talk before the exam and were encouraged to view it as a way of showing their abilities actually calmed down and did quite well. The researchers also looked at people genetically classified as "worriers" and "warriors" who had fairly high-stress jobs. A study of war veterans at the Naval Postgraduate School showed that a substantial number of expert test pilots and some Navy SEALs were homozygous for the "worrier" genes.[24] Once again we find that genes are destiny . . . sort of . . . except when they're not.

Chapter 6

AND YET ANOTHER NEW SCIENCE OF THE SAME OLD THING

ALTHOUGH THE FIELD OF SOCIOBIOLOGY BECAME A target of more and more criticism through the 1980s, the basic ideas underlying the approach remained all but impervious. A determined contingent of believers within the scientific community remained convinced that the answer to the complex questions of human behavior . . . why people do what they do . . . lies in the "hard" sciences, not in the social sciences. Like Edmund Wilson and Herbert Spencer before him, many maintain that a biological, evolutionary model will eventually subsume traditional disciplines in the social sciences and humanities. In that vein, Steven Pinker, who locates himself within the hard science end of psychology, refers to the social psychology component of his own discipline rather uncharitably as "a mishmash of interesting phenomena that are 'explained' by giving them fancy names."[1]

Pinker laments that "missing [in social psychology] is the rich deductive structure of other sciences, in which a few deep principles can generate a wealth of subtle predictions—the kind of theory that scientists praise as 'beautiful' or 'elegant.'"[2] By the 1990s, psychologists in the quest for such beauty and elegance took up the endeavors of the sociobiologists and continued their mission under a discipline with a new name, *evolutionary psychology*.

Many psychologists over the years had already gravitated toward those areas of the discipline that enjoyed the aura of science. Experimental psychology, for example, which became more prestigious than many other branches of the field, enjoyed particular esteem because of its scientific cachet.[3] The experimental approach, which gave rise to the stereotypic "rats in a maze" image, tended to focus on a search for behavioral universals, often surmising parallels across species from rodents to humans. With such a scope, differences among human cultures receive little attention.

Like their experimental colleagues, evolutionary psychologists also embrace a science model, but rather than relying on laboratory experiments with nonhuman species, they've focused on issues allegedly associated with human evolution in particular. They arrived at many of these conclusions without much study of the actual process of human evolution itself, relying instead on insights from what they've referred to as "reverse engineering."

To a great extent, evolutionary psychology has been a matter of repackaging. Its practitioners picked up the tools sociobiologists had left on the table, not to mention the old blueprints. The old assumption of genetic influence on behavior remained unshaken. Nor were evolutionary psychologists any more interested in cultural diversity than the sociobiologists before them had been, except as a source of brief rhetorical examples or "cases in point" that allegedly demonstrate human universals.

With a few notable exceptions, psychologists historically have never paid a great deal of attention to cultural differences. A brief cross-fertilization between psychology and cultural anthropology occurred in the mid-twentieth century, particularly from the 1930s through the 1970s, with explorations in fields that became known as "culture and personality" and later, "psychological anthropology."[4] Ruth Benedict, who became one of the most prominent figures in American anthropology during the first part of the twentieth century, had begun her career in the study of psychology. Most of her work focused on combining insights from both fields. But since that time, the disciplines had pretty much drifted apart.

The migration of genetic explanations from their origins in biology to a new home in psychology, via sociobiology, made some sense. Wilson and other sociobiologists had, after all, been discussing behavior all along, even though it often sounded like the behavior of genes rather than people. And who could be better suited to deal with behavior than psychologists? That's what they do. If some sort of academic matchmaking service had existed at the time, it might have found a range of compatibilities between sociobiologists and those psychologists who were drawn to the "hard science" end of the discipline.

IT BLINDED THEM WITH SCIENCE

In this new era of old ideas, we should acknowledge the difference between those who have actually been trained psychologists, with credentials, including degrees and appointments in universities or research facilities, and those who have not. Some writers who have embraced the ideas of evolutionary psychology fall into the latter category. Nonetheless, all of them have frolicked in the same general pool of ideas, operating on similar assumptions. And even those with the most distinguished credentials have often advanced conclusions that far exceed the evidence.

A key theme uniting all of them is the assumption that almost any observed human characteristics or tendencies (however accurate these observations may be in the first place) must be rooted in biological evolution. The assumption is not that the capacity for such behavior has evolved. That much is indisputable; otherwise, the behavior couldn't occur in the first place. The assumption is that genes are the cause of the behavior. The premise, generally, is that these traits are built-in, a part of our biological nature—an inherited script with the roles all set out for us, "selected for," long before we were born.

Two well-known evolutionary psychologists put it succinctly. "Because we know that the human mind is the product of the evolutionary process we know something vitally illuminating: that, aside from those properties acquired by chance, the mind consists of a set of adaptations, designed to solve the long-standing adaptive problems humans encountered as hunter-gatherers."[5] Note the "designed to" part, as opposed to "chance"? One would think that anything "designed" would require a designer. But who, or what? Humans themselves? If so, did they lose their capacity to design their lives once they began planting crops? Very odd.

A recent book on gender-related characteristics offers an example from outside the discipline. Economist Paul Seabright argues in *The War of the Sexes* that evolution has led to different life strategies among males and females, not just in the complex realm of mating but also in approaches to life in general.[6]

Males, essentially, are more prone to risk taking and to forming multiple low-commitment kinds of relationships, which gives them career advantages in the modern context. Women, however, have evolved in a setting in which they needed assistance, not only to help ensure their own survival but that of their offspring as well. As a result, according to Seabright, women are less prone to risk-taking behavior and far better at fostering deeper relationships.[7] These propensities, he argues, have had profound and generally negative implications for career advancement in today's business world. He links this to lower average income for women and other evidence of workplace disparities.

Seabright clearly takes inspiration from the work of Robert Trivers, whose work we discussed in the previous chapter.[8] Trivers argued that differential biological investment in offspring between males and females leads to greater sexual selectivity among women and a more adventurous attitude among males, because women have fewer opportunities than men to perpetuate their genes. They just can't afford to waste any. Like many such writers, Seabright is inclined to see such characteristics, however accurate or oversimplified they may be, as genetically determined rather than as the results of social, economic, or other equally biological factors such as the temporary but significant physical constraints involved in maternity.

It's interesting, in this vein, to consider the yearlong observation of "hookup culture" among young women at the University of Pennsylvania, mentioned in the last chapter. The author of the article cites author Hanna Rosin, who, "in her recent book, 'The End of Men,' argues that hooking up is a functional strategy for today's hard-charging and ambitious young women, allowing them to have enjoyable sex lives while focusing most of their energy on academic and professional goals."[9] Many blog responses appeared in response to the article, some of them in the vein of "casual sex in college? Who would have guessed?" Readers may be excused for wondering how this behavioral phenomenon might haves been "selected for" in the Pleistocene, since the previous argument sought to account for women's alleged propensity to be far more selective in their sexual activities than men.

The disparity men and women experience in career advancement, wages, and other measures in American society has been well documented. Some might suggest that, rather than looking to the Pleistocene and the innate characteristics that women supposedly acquired way back when, we might find more direct causes in more recent history and in the competing interest groups of our own society.

Another study, by credentialed psychologists, explored the idea that some widely shared fear reactions may be innate.[10] The researchers showed the test volunteers, all of them Americans, pictures of various objects—a bird, a spider, a snake, and so on—and then gauged their reactions. Among the pictures were human faces: one white, another dark. Guess which one inspired a greater fear reaction.

The researchers presented the findings to suggest an innate tendency to fear human faces that are "different" somehow. They did not suggest that dark faces are inherently scarier; they merely stressed that dark faces were different from the faces of the test subjects. The researchers did, nonetheless, present these findings as having potentially pan-human implications. "If a general preparedness to fear dissimilar others did indeed evolve, then members of another race . . . could activate such a mechanism"[11] Catch the "if . . . , then . . ." part?

Oddly, the researchers seem to have given slight attention to the history and social context of this particular country, the United States, with its dire history of racism and, as a consequence, the multiple connotations that a dark face might have as a consequence for "white" people and vice versa. If we're looking for something that supposedly evolved deep in the human past, moreover, we might remember that during the Pleistocene, when travel was on foot, it would have been extremely unlikely for humans in any given region to encounter other humans who looked very different from their own relatives. For most, that particular experience would have to have awaited the development of long-distance sea travel.

Another recent book alleges that the brains of Republicans and Democrats differ in subtle ways.[12] Given the fact that these two political parties appeared quite late in human evolution and exist only in this country, it seems difficult to determine whether, if true, this brain difference is a cause of political affiliation or a result of it. It's hard to imagine any way the gene pool could have changed much over a couple of generations—especially since the numbers of Republicans and Democrats have not only fluctuated widely during that time but have essentially reversed their geographic distributions. During the Lincoln presidency, the Republican stronghold was in the North, while Democrats prevailed in the South. After the Democrats under Presidents Kennedy and Johnson became associated with civil rights in the 1960s and 1970s, the South became strongly Republican.

We can only wish that someone had been able to examine the brains of Whigs, Know-Nothings, or Bull Moose Party members of earlier days. Still more frustrating, we have no way of knowing whether some of our Pleistocene human ancestors would gladly have become Republicans if only they'd been given the opportunity. As to an alternative explanation—the role of experience as a contributor to political affiliation? Booooring!

Many of the arguments evolutionary psychologists have advanced are serious and complex. Ultimately, however, they stand on a wobbly scaffold of assumptions. We can begin to see that rather than deep principles that are "beautiful" or "elegant" in their power to explain just about everything, what we have are ideas that are astonishingly simplistic.

The core of their arguments, of course, is the proposition that people do what they do because they have genes for that behavior. How do we know they have genes for it? Because they do it. Okay, then *why* do they have genes for it? The genes were "selected for" in the Pleistocene. How do we know that? Because people do it. It's called "reverse engineering." Oh . . . okay.

We'll discuss some of this reasoning below, including the method of "reverse engineering." At this point, let's note that these genes for behavior exist only in the realm of surmise. They remain the elusive Yetis and Sasquatches of the world of evolutionary psychology, though many claim to have seen their tracks everywhere.

Few evolutionary psychologists would express their positions in such barefaced, reductionist, and simplistic terms. Instead, they've cultivated a complex vocabulary of "modules," "evoked culture," "design features," "proximate explanations," and so on that grace their discussions with an aura of esoteric fine-point research. We might compare it to describing the oar of a rowboat as a longitudinally extended terminally spatulated lever-pivoted anthropo-manipulable aquatic propulsive mechanism with potential for duo-tandem application. (Because after all, if you want to get anywhere, it's better to pull on both oars and not just one.)

IT TAKES A VILLAGE

In 1975, the year E. O. Wilson's *Sociobiology* appeared, Steven Pinker was twenty-one years old. Born in Montreal, he did graduate work at Harvard and soon became one of the most prolific and well-known proponents of evolutionary psychology. He's been a major voice in his own right, but in a broader sense, he is also one of the latest champions of an idea that as we've seen, is actually quite old. Metaphorically, he stands on many shoulders, though not all of them belong to giants. We could see him as the child of a large, historic intellectual village of elders and ancestors who, for well over a century, have done their best to reduce human behavior to biological drives and imperatives.

The world of academics being what it is, one is likely to find allies in some places and adversaries in others. With the rise of evolutionary psychology we saw an increase in demonizing rhetoric—largely from Steven Pinker himself—directed at "behaviorists" of various sorts who questioned the assumptions of those who sought the answers to human behavior in genes. Spicing this rhetoric was a piquant sense of victimization by those "antiscience" bullies.

Pinker brings a great deal of valid scientific knowledge to the table, particularly with regard to the minute functioning of the brain and its components. He also, to his credit, wrote at one point that the evidence to date does not support the claim that IQ is biologically inherited.[13] In this, as we have seen, he departs from a long line of his predecessors and more than a few of his contemporaries. As we'll see below, however, at other points, he appears to depart from this view. On most of the standard questions that we have discussed, Pinker is a staunch proponent of biological influence on human behavior.

Despite his erudition on matters of brain science, Pinker faces the same problems that have confronted others of this intellectual tradition. They've been unable to produce any solid scientific evidence to validate their claims. In such a circumstance, the remaining strategy generally is to appeal to supporting arguments that sound perfectly reasonable. Although such a posture adopts the pose of science, it also becomes something of an art.

With regard to the old issue of whether social ranking reflects genetic merit, and whether natural processes tend to sort out those individuals with superior talents, Pinker, notwithstanding his disavowal of the inheritance of IQ about twenty-seven pages earlier, writes in the same book that "since differences in intelligence are partly inherited, and since intelligent people tend to marry other intelligent people, when a society becomes more just it will also become more stratified along genetic lines."[14]

Here, it seems, intelligence *is* partially inherited, and differentially so among various people. Pinker neglects to offer any evidence for this assertion, however. Apparently, it's just one of those things that are obvious to everybody. Pinker wrote these lines in support of Richard Herrnstein, whom we discussed in an earlier chapter on IQ. In doing so, he was responding to critics of Herrnstein's claims for the inheritance of IQ. With a lack of scientific evidence to support the assertion of inherited intelligence, we're left with a secondary tactic: an argument that the statement is so reasonable that it would be downright silly to challenge it.

Pinker often portrays his detractors (or sometimes, eminent forerunners whose views he hopes to discredit) as exasperatingly unreasonable, as slaves to irrational convictions, or worse. It can be an important advantage in any contest, of course, to choose one's adversaries. Failing that, one can caricature one's adversaries, misrepresent their views, and cherry-pick quotes out of context to the extent that the casual readers might well wonder how such fools could ever be allowed to teach anywhere. But then Pinker, despite having a distinguished faculty position at Harvard, sees college campuses as hotbeds of radical, irrational thinking anyway.[15] The scenario then becomes an epic struggle between those who are reasonable, thoughtful, and informed and the minions of ignorance, misinformation, and even mysticism.

Thus, in one of Pinker's most widely cited books, we read of behaviorists and other doubters of the wisdom of evolutionary psychology who allegedly hold the "dogma that human nature does not exist"[16] and exhibit "the mentality of a cult."[17] Many of these miscreants, it seems, share "a stated contempt for the concepts of truth, logic, and evidence."[18] As if that weren't bad enough, such a worldview, according to Pinker, can lead to the first steps on the paths to both Stalinism and Nazism.[19]

MAKING THE EXOTIC FAMILIAR, AND THE FAMILIAR GENETIC

Like most scientific endeavors, evolutionary psychology involves an attempt to discover general principles underlying what might appear to be chaotic phenomena or events in the world. Surely human behavior fits this description, for the most part. We can point to many regularities in human life, but at a more minute individual level, as any mystery novel can tell us, life is full of surprises. Émile Durkheim observed long ago that we can predict with a good deal of accuracy the frequency at which certain events will occur in a given city during the course of a year—roughly how many parking tickets, homicides, divorces, or whatever—but it would be impossible to predict the behavior of any particular individual over that span of time.[20]

Evolutionary psychology, however, the foster child of sociobiology, has persevered in the quest to identify causal principles of human behavior at the individual level by looking into our evolutionary heritage and perceiving the roots of our behavior in our evolved genetic makeup.

We can think of evolutionary psychology as a foster child of sociobiology because, in many ways, it has become separated from its origins in traditional psychology. The latter, as we noted, includes such endeavors as social psychology and methodologies that focus on child-hood, early learning, and personal histories. Evolutionary psychologists, however, rather than seeking causes of behavior early in life or in other individual experiences, have turned their gaze to the Pleistocene. Or at least, the Pleistocene as they imagine it.

Traditional psychologists, of course, have never denied a biological basis for much of human behavior. They've assumed these factors as a baseline of sorts that includes the capacity to learn, to react to experience, and to alter one's behavior accordingly. This model, of course, attributes a large part of behavior, and perhaps the most complex and interesting component, to learning. On this, most anthropologists and other social scientists would agree.

Many evolutionary psychologists, however, take issue with the idea of a broad, nondeterministic learning capacity and have argued instead for the existence of multiple "fixed action patterns,"[21] "evoked culture" patterns (or "modules," as some of them like to say),[22] or "deontic reasoning strategies" that "appear to emerge early in life."[23] Note that they "emerge," in the sense of something that was already there—waiting to be summoned from beneath the surface, perhaps.

These deontic reasoning strategies do not, apparently, develop as a result of the individual's experiences in life, but arose long ago in a distant land, from the experiences of a remote ancestor back in the Pleistocene whose reactions were "selected for" and thus became a part of our genetic heritage. By "evoked culture," environmental psychologists mean patterns that some of the uninitiated might mistake for "culture" in the conventional sense—that is, learned ideas and behavior. But no, this is genetics we're talking about.

Evolutionary psychologists do recognize learning as an important factor in human behavior, but generally they mean learning that took place back in the Pleistocene. As David Buss puts it, "Psychological mechanisms are information process devices that exist in the form they do because they have solved specific problems of survival and reproduction recurrently over human evolutionary history."[24]

Okay, but does that mean we can leave the Pleistocene behind us now and do things a little differently? Sorry, but "existing humans are necessarily designed for the previous environments of which they were a product."[25] These manifestations of "evoked culture" allegedly have specific influences over restricted domains of behavior. Examples would include such phenomena as sexual jealousy, sibling rivalry, affection toward close relatives, and so on. Evolutionary psychologists often refer to these models as powerful principles. Others might see them as the products of creative imagination.

WALTZ OF THE PSEUDOHYPOTHESES

The evolutionary psychologists' strong suit, they've argued, is that unlike social scientists . . . or even worse, humanists . . . they're doing hard science. But there's a small problem with this. As we've noted elsewhere, science is based on evidence. Typically, as we discussed earlier, the process of scientific investigation involves devising a hypothesis and testing it against data—that is, concrete information of some sort drawn from reality. A hypothesis, in other words, is a statement that potentially can be falsified. Yes, many scientific theories are extremely abstract, but ultimately, they have to do with the real world.

The excitement in physics in 2013 about the Higgs boson, which may have played a major part in the origin of the universe, comes after roughly half a century of intensive research involving thousands of scientists. Long after a physicist named Peter Higgs hypothesized its existence in 1964, researchers at the European Organization for Nuclear Research, through extensive observations based on the smashing of trillions of protons, at last concluded that the original hypothesis had likely been pretty much on target. For this, Higgs received a share of the Nobel Prize in Physics in 2013. This, of course, is a far cry from evolutionary psychologists' building elaborate conclusions on what "must have happened" in human evolution during the Pleistocene.

What if there are no data against which we could test such statements? Are they still hypotheses? Not really. Actually, they're no more than free-floating suppositions or postulates. They may be interesting. They may be outrageous. They may even be plausible. But they're not science.

Nonetheless, evolutionary psychologists have accumulated a list of specific behavior patterns that they attribute to genetic imperatives: not just old favorites such as sexual jealousy or affection for family members but also such surprising patterns as an impulse on the part of fifteen-year-olds to breast-feed (who saw that coming?) or the urge to kill one's mate.[26] Who knew? Supposedly all of these, and many, many more, had some kind of survival value in our evolutionary history. How do we know that? Because if it happens, then it must be a result of evolutionary selection, even if it only happens rarely.

How do these scholars arrive at such conclusions? Pinker explains. It's just a matter of reverse engineering.[27] Think of it this way. Instead of what we might recognize as regular engineering, such as devising a means of bringing about a result (imagine designing a steam engine to propel a boat, for example), the reverse engineers of evolutionary psychology begin with the result and then devise an explanation for how it "must have" originated. Some might call this just making stuff up.

This is hardly the stuff of science, since the only evidence or data involved lie in the observed behaviors that one wishes to explain in the first place. (And we can get into the accuracy of those observations later.) Then, simply knowing that humans have evolved from a hunting-gathering past, we can imagine conditions that must have given rise to these alleged behavior patterns in the present. These behaviors are not just phenomena calling for explanation. They function as the explanation itself, which circles back to identify its own cause. Some would call this precarious reasoning, at best. Many evolutionary psychologists would call it hard science.

But not so fast. We really don't know all that much about the details of life in the Pleistocene. Not only that, but things changed quite a bit during the course of that long period—ice ages, interstadials, pluvials, and that sort of thing. There are some things, of course, that we can deduce scientifically about the human past. We can be pretty sure, for example, that American corn, or maize, was first cultivated thousands of years ago in what is now Mexico. Why? Thanks to the work of many archaeologists and paleobotanists, we have ancient remains of the wild ancestor of maize and sequential remains at different soil levels that reveal how the domesticated plant changed over time. That's evidence. We have the same sort of information about wheat, barley, and rye in the Middle East. We don't know all we'd like to, of course, but the conclusions so far are based on concrete data.

Evolutionary psychologists can't be bothered with such details, however. The quest is not to unravel some localized, small-bore issue like the origin of a major food crop. Their ideas are far grander—pan-human even, and beyond. The empirical basis of their assertions at heart seems to be that humans evolved. That's basically it. That's what they've got. One team of well-known evolutionary psychology researchers put it very plainly for us. "All adaptations, by definition, must have a genetic basis."[28]

If, by that, they meant simply that the genetic basis for adaptation consists of the evolved capacity to learn and transmit information symbolically, thereby giving humans the capacity to adapt, this view wouldn't be especially controversial. But in fact, they assert specific genetically scripted behavior patterns—the "evoked culture" we mentioned earlier—that channel behavior in fairly narrow ways.

This term carries an impressive load of ironic contradictions in a small package. Most of us think of culture as learned. That's why there are so many different cultures in the world. But *this* culture, apparently, is "evoked" rather than learned. According to one dictionary, this would mean that it was summoned, called forth, or reawakened.[29] Pretty spooky. So the person manifesting that behavior didn't learn it. It was already there. "Evoked culture" is part of the genetic toolbox based on what some other individual long, long ago learned and passed on. But if this behavior isn't learned, why is it "culture?" Has the meaning of the word *culture* changed?

As we've noted, no gene has ever been identified as a direct cause of any sort of complex behavior. Given the way real genes actually interact among themselves and with the environment, this is not the sort of thing genes do. The plasticity and adaptive capacities of the brain itself, to the extent that one area often has the ability to compensate for an injury to another part, is also a matter of scientific knowledge. As humans, we confront genetic limitations and potentials, certainly, but usually not genetic rigidity or strict behavioral imperatives. Given the lack of evidence, why are such fabrications necessary? What do they add to our understanding?

WHAT'S WITH THE BIG BRAIN, ANYWAY?

We've seen that, despite confident claims, the scientific basis for most of the more controversial assertions of evolutionary psychologists is extremely weak. As an alternative to evidence, many have presented their views to the public as iconoclastic, heroic ideas that break through the conventional dogmas of intellectuals in the scientific establishment.

Perhaps this is an appropriate point for us to take a closer look at the non-biological-determinist view of these matters that most of the "establishment" share. Given the diversity of positions within the scientific community and the tendency for researchers to adjust their views on the basis of new evidence, we can limit ourselves to a summary of just a few issues that are pretty much matters of consensus.

To begin, it's unlikely that anyone in the social sciences would question the importance of the brain in human life, or the overwhelming evidence that the brain has evolved to its present state over millions of years—and especially over the past several hundred thousand. In terms of maintenance, the brain is the most expensive organ we have. It's huge, for one thing, compared to those of even our closest relatives—about three times as big, in some cases, as that of an average chimpanzee.

The brain requires a continuous and copious flow of blood to keep it from shutting down. Its temperature is so important that the body will allow other parts, such as extremities and limbs, to freeze before allowing the brain to cool. Overheating can also cause it to malfunction, as anyone who's had a high fever can attest. And, of course, the brain is highly vulnerable to physical injury. A whack on the head with a degree of force that might only cause a nasty welt on the buttocks can put the brain out of commission. We can continue to function with cracked ribs but not usually with a cracked skull.

As sociobiologists and evolutionary psychiatrists have often pointed out, human reproduction involves a long gestation period and an even longer period of infant helplessness. Trivers and others have pointed to this as a basis for their assertions of sexual differences in parental investment. Why does it take so long? Might as well blame that on the brain too.

The human brain has become so large that it often has trouble passing through the birth canal—a fact that has made childbirth much more dangerous for humans than for other species. As an evolutionary adaptation of sorts, human infants are born at a stage of development that in other species would be embryonic. This means that human babies have a much longer period of dependency and need for care than other species. A puppy at the age of six months is essentially a young adult dog. A child of the same age, as we know, has a long way to go before reaching a comparable stage.

Of course, sociobiologists and evolutionary psychologists might argue, this is a reflection of the importance of the brain in directing our evolved behavior patterns. But wait a minute. The smaller, more compact, and let's face it, simpler brains of other species obviously do that for them pretty well, without all the costs. Not only that, but their brains even seem to allow some adaptive behavioral flexibility—what we like to call learning.

A dog doesn't need to be taught how to chase cars; yet, there were no cars around during the course of canine evolution. But there were certainly other things to chase; so "chasing stuff" certainly has deep roots in the brain of a dog. Yet most dogs don't just chase everything that moves. Most dogs (though perhaps not all) are capable of learning not to chase cars. Even at this level of brain complexity, then, despite ample innate guides to behavior, a dog still has some flexibility for learning outside the realm of reflexes and instincts.

Since there's no doubt that the human brain has lots of downsides, we might ask ourselves why it evolved into such a monstrosity. If guiding behavior is its main function, why are humans saddled with such an expensive, cumbersome organ when a much smaller one would do nicely? If we choose to think in terms of cost-benefit analysis, the tremendous cost of the brain must have had offsetting benefits, must have provided advantages that far outweighed what a smaller brain could have offered. What selective advantages, as evolutionary psychologists are wont to ask, could have led to this? The obvious answer is that human brains have important functions over and beyond those of dogs. Humans rely much more on learning for survival.

An important consequence of an enhanced ability to learn is the capacity for flexibility in behavior, which is crucial to adaptability. Dogs can learn, as we've noted. Chimps can learn even more. But none of them holds a candle to the average human being in terms of learning capacity. And why would we have developed such a capacity, at such great biological cost, if the more "primitive" parts of the brain were still going to tell us what to do anyway?

Clearly, there's more to it than that. It's not just that the human brain has more storage capacity for information, like some sort of a knowledge silo. It also gives us an ability that no other species has, as far as we know: language. As we noted in the last chapter, language is, indeed, a result of the physical, evolved nature of the organ. We know that injuries to certain areas of the brain can impair language ability in very specific ways. And we also know that the crucial parts of the brain that have allowed humans to develop language don't exist in the brains of chimpanzees or any of the other primates.

Okay, so we have language. We can talk to one another. But that's just a small part of it. We talk to one another in lots of different languages. We're "hard-wired," as some sociobiologists used to say, to learn some language or other. We can't help it, unless we suffer from some disability. But we're not wired to learn any particular language. The potential for linguistic diversity, as we know, is vast. Not only does everyone have a language, but lots of people are fluent in three, four, or even more different, mutually unintelligible languages.

We know that language functions at various levels. The lowest of these, perhaps, is to convey something about the immediate situation. Even a dog can do that, by nonlinguistic means. Anyone who's had a dog has no problem understanding most of a dog's basic messages: "I'm glad to see you." "I'm sad." "I'm hungry." "I want that cat off my lawn, now!"

But we also know that language is far richer than that. It conveys more than information about the here and now. People use language to tell lies, share fantasies, perpetuate myths, recount history, and specu-late about the future. It acts as a vehicle for thoughts that arise in the brain and remain there as memory. In the midst of these complex pro-cesses that we undoubtedly are able to carry out because of the course of human evolution, we continually learn countless things from one an-other, including how to behave.

All this ties in with evolved brain size and complexity but also with the prolonged infancy of humans. That slow road to maturity requires a lengthy period of physical care and nurturance and, just as important, a long and continuous process of socialization. People become members of their group, knowing how to behave appropriately and generally even wanting to, because they're deeply affected by what we sometimes call their "formative" years. They often behave quite differently from people in another group—not because of a different set of genes, but because the experiences of those formative years often differ profoundly among human societies.

All of this, presumably, would not provoke much dissent among social scientists. Indeed, it amounts to conventional wisdom. What justification, then, is there for asserting that despite our having evolved with such a brain, genes or genetic modules tell us what to do?

WHAT, INDEED?

The genes or genetic modules that evolutionary psychol-ogists claim direct our numerous complex patterns of behavior seem to be imagined clumps of molecules in combinations "selected for" by some circumstances back in the Pleistocene. We've noted some of the prob-lems with this idea, including the fact that almost nothing about hu-mans, except for things like eye color or blood type, is a direct result of the presence or absence of a particular gene. Almost every aspect of the human body is the result of an interaction among numerous genes, and their interactions, in turn, with other nongenetic (epigenetic) factors. This is true of stature, body weight, the manifestation of many diseases, and so on. How much more complex would genetic interactions have to be even if, as many argue, there were instructions somewhere in our DNA for such things as sharing with one's close kin, sexual jealousy, or aggression?

To address this at a simpler level, let's look again at something universally human: walking. The ability—in fact, the propensity—of humans to walk is certainly a product of evolution. It occurred millions of years ago in our earliest ancestors. Bipedal locomotion, in fact, is considered a diagnostic feature in distinguishing the early appearance of the hominin line. That's our thing. We can't fly; we don't even run very fast. And crawling gets really uncomfortable after awhile. A milestone in many families is when a baby takes her first steps.

We take all of this for granted, unless some injury, illness, or old age interferes with this ability. But let's consider what's involved. Even though we can agree that walking has a genetic basis, to suggest that we have a "gene for walking" would be ridiculous. Let's begin with the complex anatomy of the foot, with its specialized bone structure—a heel to absorb impact and tarsal and multiple small bones of the arch of the foot to act as shock absorbers; the sturdy, articulated long bones of the legs culminating in the ball and socket joints up under the hip to bear the weight of the torso; the complex muscular structures involving the gluteal muscles of that distinct hallmark of humanity, which some might rudely refer to as the "butt"; the semicircular canals of the inner ear to assist in keeping balance while tottering up on two limbs.

There's a lot going on here. But those are just some of the complex mechanisms involved in walking around. The number of genes implicated, their interactions, and the play of such epigenetic factors such as nutrition and continual use to allow muscle and bone development are considerable. And we haven't even talked about neural connections and blood circulation. Then we can get into such cultural things as footwear—is it sandals, sneakers, spike heels, moccasins, foot-binding, or just bare feet? The propensity to develop foot calluses from walking around barefoot is another aspect of a genetic potential responding to nongenetic conditions. Wearers of shoes are more likely to keep the soles of their feet soft and delicate.

Okay, so we have the complex genetic and epigenetic aspects of the evolved human ability to walk. How much does that tell us about genetic, innate influences on human behavior? Not much. "Walk a mile in my shoes." "It was a walk in the park. "Get off my lawn!" "She walked all over him." The contexts, meanings, and reasons for the ways in which this evolved biological propensity plays a part in human life far exceeds what any genetic modules could account for.

But what about some of the behavioral modules that evolutionary psychologists have alleged? We've been told that evolution has left us with built-in propensities for male promiscuity, sexual jealousy, aggression, and sharing with one's closest siblings. (Many a parent with small children might wish there were such a thing as a gene for sharing with siblings.)

Again, there's no doubt that humans share a genetic capacity for all of these forms of behavior, since they happen. But they don't happen all the time. And when they do, we can usually attribute them to specific factors or circumstances, given enough information. But what about the many times when they don't manifest themselves? Is this because they've been suppressed somehow? Are the individuals who fail to follow the genetic script somehow deficient in the DNA department? Or is there, in fact, no such gene in the first place?

Let's take, for example, the "module" that leads to male profligacy. How do we handle the fact that most men—depending on their circumstances, of course—don't have sex frequently with lots of different women? One could respond that they probably would like to—they just aren't that lucky, or attractive, or whatever. But how do we know that women don't feel the same way? A number of recent writings suggest that maybe, just maybe, they do.

It might seem odd to someone who's not from around here that given the fact that women outnumber men in the population their feelings on this matter should be so unclear. In many cultures, this would pose less of a mystery. One might almost suspect that either no one's asking, or nobody's talking.

But wait—even in our own culture, isn't there a large body of literature, much of it by women, that has a great deal to tell us about women's perspectives on this issue? Not to mention recent discussions based on more formal scientific sources.[30]

To substantiate the assertion of a genetic drive toward male profligacy and female judicious selectivity, we would have to determine the answer to these questions.

As we discussed earlier, American women participating in one research project turned out to be more than equal to men in their number of sexual partners. The women just tended—for cultural reasons, no doubt—to underreport this information until they were told that they were attached to a lie detector. And from what we know of American society, it's easy to suspect that men may well have overreported their own sexual adventures.

What good does it do to posit genetic modules that allegedly compel certain kinds of behavior when in many cases they clearly don't, since the behavior doesn't occur? Is it that these modules are supposedly overridden by learned cultural values or social rules? Or that they somehow go dormant at times? If so, then how important could they be? Are they merely "tendencies?" Or, given the fact that there's no scientific evidence for their existence, why should we suppose that they're anything more than fantasy?

Chapter 7

That's Just about Enough of That

IN MAKING THE CASE FOR BIOLOGICAL EXPLANATION, Steven Pinker and others have characterized alternative approaches in less than complimentary terms. Given the intensity of their comments, perhaps it's appropriate to give these points some serious consideration.

Pinker states, for example, that his intellectual adversaries "deny human nature."[1] On this point, as on so many others, it all depends. It seems doubtful that any reputable scholar would deny that we humans share certain qualities and characteristics that distinguish us from other creatures, or that in fact, the "nature" of humans is profoundly distinct. On the other hand, this doesn't seem to be quite what Pinker and the other evolutionary psychologists have in mind when they use the term. Perhaps we could say that the emphasis is more on the "nature" than the "human" part, and they apparently see less distinction between the two than many scholars might.

To the extent that they have in mind a species whose actions, decisions, and perspectives result primarily from evolved quasi-instincts or drives, then, no. That version of behavioral causality is not part of the explanatory tool kit of most people who study humanity. To many such scholars the biologically driven model, despite the obfuscating rhetoric and hyper-technical terminology that often accompanies it, is far too simplistic.

Pinker accuses social scientists of subscribing to the quaint Enlightenment-era idea of the human mind as a "blank slate." He's so taken with this idea that he used it for the title of one of his most important books. This concept—the *tabula rasa*, as we used to call it—is, of course, one of the foundational ideas of the modern social sciences. Its significance during the Enlightenment was to rebut both theological and racist ideas of the time. Essentially, it was an assertion of fundamental human equality of potential. At a basic level it implies that to a great extent, who we are is a result of what we learn. As Susan McKinnon and Sydel Silverman put it, "It is also in the nature of humans to be fundamentally and intensely social and to exist—always and everywhere—within networks of social relations and webs of cultural meanings that they both shape and are shaped by."[2]

Some would argue that this idea has stood up well over time. It's a bit difficult to see this archaic but once useful idea as pernicious a threat to our well-being as Pinker does. Although we associate it with the Enlightenment, it's actually been around since Aristotle. So far we've survived. Sounding the alarm, though, Pinker warns us that the idea of the blank slate "is an anti-life, anti-human theoretical abstraction that denies our common humanity, our inherent interests, and our individual preferences."[3] Ironically, many thinkers over the years have seen the concept as an implicit assertion of our common humanity. But at least he got the "abstraction" part right.

Pinker seems particularly exasperated with anthropologists, who come across as archenemies of the acceptance of "human nature." So extreme are some of them in denying a common human nature, he sputters, that "a few anthropologists say there are cultures with no emotions at all!"[4]

Really? Can that be? Let's take a look at the sources he cites. One is a collection of works in which the anthropologist Richard Shweder discusses cultural variations in the ways people categorize different states of feeling.[5] No emotions? Wait a minute. That's not what Shweder is saying.

Shweder's point is not that the people in this particular culture have no emotions. He discusses a range of emotional states that they recognize. But the terms they use to refer to these emotions do not correspond directly to emotional terms that Americans would use. They are not, in other words, directly translatable. Of course, even among Western cultures, we incorporate some foreign terms for emotions into our own language because we don't really have a word that means exactly the same thing. What's the English word for *ennui*? "Boredom" doesn't quite capture it. The term *Schadenfreude* also seems to pop up in many conversations these days. Likewise, many a busy Italian knows what *lo stress* feels like.

Shweder notes that "in fact, there are NO emotion terms which can be matched neatly across language and culture boundaries . . . there are NO universal emotion concepts, lexicalized in all languages of the world."[6] Interesting observation, but not nearly as bizarre as we might have expected from Pinker's outrage. Not only that, but unlike many of Pinker's assertions, Shweder's observations are directly testable.

Not that humans, either as a species or as individuals, lack any in-born proclivities or innate limits to our abilities, or that we don't differ in our potentials. But to push learning to the background or see it as superficial dressing on a range of evolved modules' directions for specific forms of behavior, despite the science-like presentation of these assertions, seems to push things back to a pre-Enlightenment, almost medieval way of thinking. It's almost reminiscent of the "vital essences" of earlier times.

"WHEN WILD IN THE WOODS THE NOBLE SAVAGE RAN"[7]

Pinker is also fond of accusing social scientists of adhering to the idea of the "noble savage," which assumes a fundamental goodness in human nature. Where do we begin here? Should we start with the insulting, almost racist quality of exhuming this centuries-old term? Noble, okay. But most of us don't talk much about savages anymore. Or should we address the audacity of this argument coming from one who shows no signs of ever having experienced immersion in, or even having any deep familiarity with, any non-Western population? His acquaintance with the subject is amply clear in his statement that "traditional peoples believe in sympathetic magic, otherwise known as voodoo."[8]

But let's put that aside, take a breath, and have a look at the origins of this phrase in the seventeenth century. It was a time when European ignorance of peoples in other parts of the world was profound. The term *noble savage* embodies a caricature that social scientists over the past century or so have worked assiduously to dispel—not least because it was mostly a figment of European daydreaming.

One of Pinker's main targets is cultural anthropology, especially the American version typified by those who were influenced by Franz Boas. We might remember Boas from chapter 2 dealing with eugenics, with his head measurements of immigrants. It's somewhat ironic that Pinker should take Boasians to task for allegedly denying the biological nature of humans. Boas established American anthropology, as distinct from the European version, as a "four fields" discipline incorporating not only the global study of contemporary cultures, language, and archaeology but also human biology, including evolution.

Pinker acknowledges that Boas had some good points, although his major accolade—that Boas considered the cultures of Europe to be superior to others—is entirely at odds with the thrust of Boas's work.[9] In general, though, in Pinker's view the poor old fellow was somewhat misguided, and his followers seem to have amplified his errors. The noble savage doctrine, as Pinker refers to it, is one of those.

Here again we find ourselves ankle-deep in irony. One of Boas's major contributions, many would agree, was to approach the variety of human cultures as alternative and essentially, at a fundamental level, equivalent ways of living. His focus was on the complexities, the meanings, and the historical "hows" and "whys" that might help us understand the lives of people in diverse societies. But there was no assumption that people in small communities were necessarily any more or less "noble" than anyone else. Indeed, it would be hard to find the word *noble* in any of Boas's voluminous writings.

More importantly, perhaps, it would be fair to say that a major thrust of Boasian anthropology was to eradicate the idea of the savage, with all of its connotations of ignorance, primitiveness, and—well, savagery— that had attached to it over previous centuries. Boasians didn't hold some idiotic view of naked noblemen luxuriating in a sylvan glade. These researchers actually lived with people, shared their food, and talked to them about their views of things. They described complex human beings dealing with universal human issues and problems in their own ways.

This also leads to the issue of relativism, another favorite target of Pinker's. Whether Pinker recognizes it nor not, though, relativism comes in several versions. Some writers confuse cultural relativism—which Boasians espoused, and most anthropologists still do—with moral relativism.

Cultural relativism means that when attempting to understand a particular custom, even one that an outside observer might find repugnant, the question is *why* people do it. Simply to dismiss it as repulsive and let it go at that doesn't really accomplish anything. It's really little more than self-indulgence. To get to the why, one needs to set revulsion aside and ask other questions. To put it another way, cultural relativism allows us to move beyond making facile judgments about whether people should or shouldn't do something and try to understand the underlying reasons for why they do it.

Pinker bemoans the alleged infection of college campuses with "cultural relativism" to the extent that students are no longer willing to make moral judgments.[10] But that's not cultural relativism. That's moral relativism—something that Boas certainly did not embrace. During his long professional life, Boas engaged with numerous social issues and tried to bring about change, fighting for civil rights and against racism in particular.

Boas was an outspoken opponent of racial discrimination and the eugenics movement. He joined the struggle for fair treatment of the Scottsboro Boys, nine African American teenagers in Alabama who in 1931were sentenced to hanging, after an obviously unfair trial, for allegedly raping two "white" women. Several of his students also were significant figures in the civil rights movement. Ruth Benedict and her colleague Gene Weltfish wrote a pamphlet *The Races of Mankind* arguing the fallacy of "racial" categories for distribution to U.S. troops.[11] Melville Herskovits carried out studies to contradict the assumption that slaves had lost any traces of African culture and were therefore "cultureless" except for what they had acquired in America.[12] Zora Neale Hurston, another student of Boas's, was a major figure in the Harlem Renaissance.[13] Boas was not a moral relativist, nor were his students.

But in trying to understand such issues as, for example, infanticide, head-hunting, or vengeance killings in other cultures, a simple conclusion that "they shouldn't do that" doesn't take us very far. The attempt to discover what underlies these patterns, the conditions that may have given rise to them and perpetuated them, and the beliefs surrounding them, is a very different endeavor.

STEVEN—YOU LOOK AS IF YOU'VE SEEN A GHOST

In his outrage at scholars who allegedly deny human nature, Pinker accuses them of subscribing to something he calls "the ghost in the machine."[14] What he means by that, apparently, is some sort of nonhuman guiding force that causes things to happen in human affairs. Although many anthropologists have studied and tried to understand the beliefs of people in societies that actually involve ancestral ghosts, animal spirits, and whatnot, it's safe to say that few of them believe in ghosts. Many are also atheists—perhaps because after exploring multiple belief systems involving the supernatural, it can be difficult to maintain a conviction that one of them, in particular, is correct.

But what Pinker seems to be referring to is not ghosts in the conventional scary movie sense, but the concept of culture—or *his* concept of *anthropologists'* concept of culture. Supposedly, in his view, anthropologists imagine a supra-individual entity that controls minds and causes things to happen above and beyond the wills of individuals. This offers us still another dose of irony since, as we've seen, sociobiologists and evolutionary psychologists are fond of talking about genes and evolved modules in just that way.

In any case, as far as "culture" is concerned, the development of the concept by Boasians a century or more ago was basically a rebuttal to the racist thinking of the times. Human behavioral characteristics, whatever they happened to be, were results of learning rather than inborn "racial" traits. Sound familiar? Culture focused on learning as a result of being part of a collective entity, a society.

The idea was not that society, or culture, had a mind of its own or some sort of autonomous volition. It did, however, involve the idea of supra-individual phenomena: shared beliefs, "culture patterns," and the like. If a significant number of people share a particular idea and pass it on to their children—the idea that owls are messengers from the dead, for example, or that putting food into the mouth with the left hand is repulsive—we can say that even if that idea exists only in people's minds, it has an existence of sorts over and above any of them as individuals. If one or more of them should die or decide that the idea is nonsense, the idea nonetheless continues to exist without them, perpetuated by other people. Until, of course, a time comes when no one at all holds the belief anymore. (Although even then, if the idea was recorded or remembered, it would become part of history, continuing to exist even without adherents.)

CALM DOWN; IT'S ONLY AN ABSTRACTION

Alfred Kroeber, one of Boas's more prominent students, devised a concept to express the supra-individual aspects of collective belief and knowledge. He referred to this as the "superorganic."[15]

Kroeber's model of the superorganic was not without its critics even in the early twentieth century when he developed it. But its thrust was to offer a way of thinking about the human experience that emphasized the collective nature of belief systems, which often have trajectories of their own, beyond the control of any particular individuals. Economists use the term *macro* for overall, long-term collective phenomena. An important aspect of the superorganic way of looking at human behavior is that it operated according to dynamics of its own, over and above any genetic, biological, or "racial" imperatives.

This is not Pinker's cup of tea, of course. Nor, after more than half a century, is it a model to which most anthropologists nowadays would subscribe in full. It did not, however, involve anything that might summon the rather bizarre image of Pinker's worst recurring nightmare, the Ghost in the Machine.

In considering the long endeavors of anthropologists, historians, and other non-biological-determinists to make some sense of the human experience—some more successful than others, of course—one can't help being struck by the complexities of the subject. It should be unnecessary to point out that humans and their history are complicated—especially when we factor in not only their interactions with one another, but with the range of environmental factors and conditions in which they've operated and with which they've interacted. Pinker's accounts of human life, however, seem refreshingly simple. Perhaps this is one reason for their appeal in some circles.

Despite his dauntingly detailed discussions of the physical nature of the brain as an organ,[16] his descriptions of human life sometimes come across as reports sent by a Martian to the home base describing the inhabitants of a strange planet:

> Reciprocators who help others who have helped them and who shun or punish others who have failed to help them, will enjoy the benefits of gains in trade and outcompete individualists, cheaters, and pure altruists . . . they remember each other as individuals (perhaps with the help of dedicated regions of the brain), and have an eagle eye and a flypaper memory for cheaters. They feel moralistic emotions—liking, sympathy, gratitude, guilt, shame, and anger—that are uncanny implementations of the strategies for reciprocal altruism in computer simulations and mathematical models.[17]

At other times, his analyses sound almost like descriptions of a video game: "If an obstacle stands in the way of something an organism needs, it should neutralize the obstacle by disabling or eliminating it. This includes obstacles that happen to be other human beings—say, ones that are monopolizing desirable land or sources of food."[18]

Perhaps this is a result of emphasizing the "genetic strategies" and modules of "evolved culture" that have been "selected for" to "enhance the perpetuation" of genes. In these descriptions, we can see some resemblance to human life as we know it, but it might well be a simple dramatization in which the actors are robots. Even paramecia swimming chaotically in a drop of pond water under a microscope seem to have more discretionary options available to them than the genetically programmed humans of the evolutionary psychologist's world.

ARE YOU STILL HERE?

Is there, at long last, no end to this? It might be useful at this point to take a long view of the process and see that in fact, the primary "survival strategy" of this perennial endeavor to establish a biological basis for human behavior has been its dogged persistence, rather than any scientific validation. After over a century, despite tremendous progress in all of the sciences, we're left with no conclusive evidence to support it.

The early versions of raw, unabashed racist and classist doctrines of the nineteenth and twentieth centuries lost favor a long time ago in most of the scientific community, although as we know, they haven't vanished entirely. As we've seen in the United States in the past few years, they seem to lie deep within the very fabric of American culture, ready to bloom like toxic mold when the conditions are right. But they no longer have a place in mainstream science. The quest for a genetic basis for human behavior in general, however, continues unabashed among true believers.

Despite scientific advances, including major breakthroughs in genetics, no genes that shape what we might call "normal" behavior have ever turned up. There clearly are some genes (or the lack of some genes) that can produce pathologies, but these are aberrations, genetic glitches, not general guides for the behavior of the species.

We've learned enough about the action of genes to know that single genes usually don't do very much on their own.[19] Usually, they interact with other genes, if they act at all. Often they don't seem to do much of anything. When they do manifest themselves, whether alone or in combinations, they do so in various ways in reaction to different circumstances. We call these epigenetic phenomena, that is, phenomena that result from the interaction of genes and other factors.[20] Even the recent advances in mapping the human genome haven't changed this bottom line, except perhaps to underscore it.

The discovery that apparently we only have about twenty thousand genes—of which only about 1.6 percent differ from the genome of chimpanzees—reinforces this.[21] It doesn't leave a whole lot of genes to carry out the multiple highly specific directives that evolutionary psychologists attribute to them. It underlines the fact that many genes have multiple functions (pleiotropy), that their combined interactions are many and important, and that epigenetic factors play a crucial role. Given that, the idea of "a gene for sharing with siblings" or "a gene for sexual jealousy" is downright silly. Advances in the science of genetics have torn gaping holes in the "hard science" lab coat of evolutionary psychology, in some embarrassing places.

That leaves us with other questions. Why has the quest for genetic explanations of behavior been so persistent? What's the appeal? Aside from the space program and major discoveries in medicine, it's difficult to find any scientific enterprise that has generated more public interest and even excitement in the media than purported "discoveries" that some aspect of human behavior is genetically determined.

It could also be that most people, if they think much about it at all, don't necessarily buy into any such model of biological determinism. Most people, perhaps, really do feel that their actions are pretty much up to them. "Scientific" claims of biological imperatives, therefore, may be somewhat surprising and even counterintuitive, but for most people, they may not rise above the level of mildly entertaining ideas. The news media tend to focus on the unusual, the unexpected. We don't read headlines about cars that didn't crash. It could be, then, that neither the public in general nor the news media wait avidly for the next confirmation of biologically evolved behavioral traits. More often, perhaps, they greet these stories with some surprise, skepticism, and, in some cases, amusement, and then go on with their lives.

There is another possibility, though. It might be that in some quarters this is welcome news. It can seem to simplify things a bit, in a world where the complexities of human behavior often seem daunting. What a relief to have a clear, straightforward biological reason for something . . . even if that something might be male philandering, social conflict, or worse. During a discussion of war on a National Public Radio talk show called *On Point* in May 2013, a caller explained that war is inevitable because some people have "the warrior gene." If some of these things were a matter of biological imperatives, they might still be bad news, but not something we can do much about, except maybe to try and suppress these "natural urges" and keep them under control.

WHAT'S THE BIG DEAL?

With all the academic debate, what difference does it really make to most people, after all? To what extent do heated arguments among professors, perhaps even creative name-calling, matter to people going about their business in the rest of society?

Actually, they matter a lot. They have implications for the support of public programs, for the way we approach various social issues, for the way we define problems in the first place. Sometimes they only provide talking points for officials who have already made up their minds, but they still provide ammunition of sorts in important debates. It can be handy for one side or another to state that "studies show . . ."

Can we improve the human condition substantially through attempts to address social problems, by investing in education and taking other initiatives? Or is everyone basically just acting out a Paleolithic script with predetermined roles of winners and losers, built-in gender inequality, and innate aggression that somehow evolved because it was "selected for" early in the Pleistocene, yet inexplicably stopped evolving long ago and became "fixed" from that time on?

As we've seen, ideas of biological fixity have been around for a long time, largely impervious to the lack of scientific evidence to support them. In the twenty-first century, they continue to permeate the cultural substrate, and in some quarters, they seem to have as much appeal as ever. The Pleistocene continues to fascinate—albeit often as a cartoonish image. Commonplace sexual attraction has now been diagnosed as a perception of "good genes" in a prospective partner—a "carryover from the Paleolithic"—even if procreation may be the last thing on the minds of the people about to get busy, except perhaps, how they can make sure conception doesn't happen.

In recent years, the "Paleolithic diet," in various versions, has stirred some interest, even though the specified foods often have little to do with the way our ancestors would have subsisted back in the day.[22] And what guy doesn't want a "man cave," a den with an authentic neo-Paleolithic flat-screen TV where he can watch guys with "the warrior gene" make spectacular tackles and broken-field runs—not to mention a nice bar with a food cache and a fridge full of beer? It doesn't get more "selected for" than that.

More disturbingly, age-old arguments against investment in social programs for those who suffer from a range of social and economic disadvantages are still very much in the air. We might recall the flurry of dubious IQ arguments that arose in the 1960s in reaction to the establishment of the Head Start program. In March 2013, a dysfunctional Congress froze much of the federal budget as part of a national fiasco known as "the sequester." One of the consequences was the gutting of Head Start funding.

The move caused a good deal of consternation and anger, especially since disparity in the distribution of wealth in the United States had sharply widened in the previous decade, and children in poverty had more need than ever for such assistance.[23] Government studies, moreover, had demonstrated that even though Head Start was not enough to overcome such factors as homelessness, street violence, inadequate nutrition, and other conditions of life in impoverished areas, the program had nonetheless had a consistent positive effect over the years.[24]

Yet in March 2013, one comedian and iconoclast who has often favored liberal causes announced flatly during a discussion with his guests on his television show that "Head Start doesn't work." Being a professional entertainer, of course, he was under no compunction to cite evidence. Usually, in such a context, it's more than sufficient for the entertainer to stare intently with a long face and brows raised slightly to the "I'm sincere" position. If he can add "studies have shown," it's even more powerful. Except that, in this case, as we've seen, studies have shown something quite different.

We have to ask what might have brought about this individual's conclusion about Head Start? We can only speculate, of course, although we can probably rule out the likelihood that he had conducted a personal study of the subject. More likely, his opinions, like those of most of us on most topics, arose from something he'd read in the media, or something he'd heard somewhere. And this is the problem.

Of course, the media are free to publish anything they want, and, in the broadcast realm, to say almost anything. We've seen the limits tested severely many times in the past few years. If a media outlet consistently, or even frequently, spews nonsense, then consumers may draw their own conclusions about the quality of the source. This is relatively easy when it comes to such issues as the president's birth certificate or some alleged secret plot by terrorists to impose Sharia law in Kansas. It can be more confusing, and misleading, when much of the misinformation comes from sources with impressive scientific credentials.

We've heard a long, discordant chorus of such voices through the past century. Remember Konrad Lorenz, recipient of the Nobel Prize for his work with geese and fish? His take on human aggression didn't stand up as well to critical scrutiny. And then there was William Shockley, Nobel laureate for his work on the transistor. He was also a proud, vocal racist whose espousal of eugenics did not, fortunately, have as great an impact on American society as his technological breakthrough.

In these cases, it was fairly easy for most people to sort out the nonsense, although the receptive audience was still larger than some of us might have hoped. The ideas these figures espoused have been resilient. These fleeting stars of biological determinism were bearers of a long tradition. One could imagine an ideological relay race, with someone always ready to pick up the baton and run the next lap.

WHAT'S THE SCORE SO FAR?

We've now undergone well over a century of attempts to impose a biological explanatory framework on human behavior. During that era, what scientific achievements have occurred? We've gone from horse-drawn vehicles to space travel, from written letters carried on foot or horseback to multipurpose cell phones and instant texting over a global range. We've seen the development of the polio vaccine, the eradication of smallpox, MRI diagnostics, and laser surgery in medicine, the theory of relativity, quantum physics, and the confirmation of the Higgs boson. So how's that genetic/behavior thing going?

The basic premise goes like this: Humans are part of nature; therefore, most aspects of human life are the direct results of our biology. Back in the day, the issue was about differences among groups or populations—social classes, nationalities, and "races." Many scholars, writers, and political figures assumed that perceived differences among these various categories of people were aspects of groups' biological "nature"; that it was "in the blood." But eventually some of these terms and ideas became less fashionable and even a bit embarrassing to some. Going beyond "racial" differences, some suggested instead that humans *in general* share some innate, evolved characteristics from our prehistoric heritage: a killer instinct, a territorial imperative, an innate tendency for aggression, and so on.

In time that also came to seem a bit crude and overblown, not to mention its shaky scientific basis. On the other hand, once we had IQ testing, by the 1960s we could attribute differences in intelligence—or more precisely, differences in IQ test scores, which many writers treated as the same thing—to inherited biological traits. Even "racial" differences made a comeback in scholarly publications. That provided some talking points for opponents of social programs such as Head Start. Still, there were plenty of skeptics with regard to that position, especially since once again, there didn't seem to be a great deal of solid scientific evidence to support it. Remember the posthumous undoing of Sir Cyril Burt, prominent authority on IQ tests for identical twins?

In the 1970s, sociobiology arrived to pick up the baton. Here, the search for biological imperatives turned back to the animal kingdom, not just mammals this time, but a far wider range of species, including social insects. What the whole game of life is about, sociobiologists informed us, is the need to pass on as many of one's genes as possible. Hence, the attention shifted away from populations, which had been the major focus of biology to that point. It was now about the ability of individuals within a population to transmit genes in competition with other members of the same population. Indeed, it almost seemed as though the genes were in charge, strategizing and calling the shots.

Sociobiology stressed competition and conflict, since not all genes could succeed. That, after all, was what selection was all about.

Eventually this idea saw further embellishment. The strategies of males and females for passing on as many genes as possible had to differ as well, since the biology of reproduction with long gestation periods, nursing, and all that puts women at a disadvantage when it comes to the gene transmission contest. And passing on genes, let's not forget, is what it's all about. These factors, of course, would have lots of implications for gender relations (which are built-in, "naturally," right?).

When the spotlight on sociobiology eventually dimmed, largely as a result of growing criticism from the audience and a dearth of empirical evidence, adherents to the cause picked up the baton once again and carried on under the banner of evolutionary psychology. The franchise lost some players, recruited others, and continued the game with a new team logo. They currently hold the field.

After all of that, involving generations of researchers engaged in the same quest, what progress have we seen?

The study of human evolution during this time has seen many advances. We now have a map of the human genome, and the overall field of human genetics has progressed far beyond where it was even a couple of decades ago. Has any of this helped to bolster the biological/behavioral position?

No. The primary advances in what is now evolutionary psychology have involved a certain amount of shape-shifting and adopting protective coloration to fit the times. But increased insights in the relevant fields of biological science have, in fact, undermined many of evolutionary psychology's central positions. The more we've learned about genetics, the less plausible they've become.

SOME THINGS WE DO KNOW ABOUT THE PLEISTOCENE

Evolutionary psychologists, like sociobiologists before them, have based their assertions on a range of suppositions that we can best address one at a time. Many of the most important suppositions involve alleged human adaptations that arose during the Pliocene and

Pleistocene geological eras, which cover the last several million years of human evolution and, by most estimates, ended about ten thousand years ago with the close of the last ice age. Other suppositions arise from comparisons of humans with other primates, with a range of assumptions drawn from them, assumptions about what researchers consider basic, universal gender roles, and in a more general sense, claims to a stronger grounding in "hard science" than other fields of study that address the human experience. Many of these we've briefly addressed earlier, but they deserve more attention. Let's begin with the Pleistocene.

Evolutionary psychologists propose an array of genetic modules that allegedly compel certain kinds of behavior. These patterns of behavior, they argue, arose earlier in human evolution because the conditions of the Pleistocene "selected for" them. Hence, these behavior modules became firmly entrenched in our genome, and they continue to shape our behavior in the present. These include the following examples:[25] "anti free-ride adaptations in cooperative groups";[26] "cheater-detection adaptations in social exchange";[27] "female superiority in spatial location memory as part of a gathering adaptation";[28] "functional attributes of male and female short-term mating strategies";[29] and "sex-differentiated deception tactics in human mating."[30]

The genetic bases for these rather complex behavioral phenomena, according to the authors of the article cited, have been "confirmed by multiple methods in multiple samples by multiple investigators."[31] Others, they say, have yet to be as thoroughly tested.

The Pleistocene explanation assumes a number of things, one of the most important of which is that this geological era, the context to which early humans had to adapt, was a relatively constant factor. Otherwise, how could such stable, and all but inexorable, genetic modules have persisted through these eons? Evolutionary psychologists don't go into much detail in discussing the Pleistocene environment, except for conveying the impression that it must have been pretty rough.

But the thing is, we're talking about millions of square miles, from well north of the Tropic of Cancer to well south of the Tropic of Capricorn, over a couple of million years. And that's just Africa. During the latter part of the Pleistocene, let's not forget, hominins had also ventured into Europe and Asia. How many different kinds of environmental contexts did that involve? Rain forests? Plains? Savannahs? Mountains? Lakeshores? Seacoasts? Deserts? And how many kinds of food sources, involving how many different kinds of hunting, fishing, and gathering strategies to procure and distribute them? Through how many kinds of social arrangements and organizations, divisions of labor, and redistributive mechanisms?

We don't know, of course. But we can be sure that the human past, in a panoramic sense over time and place, does not lend itself to a one-size-fits-all model. And surely it goes far beyond the simplistic competition for food and procreation scenario that evolutionary psychologists imagine.

But the setting gets more complicated than that. Even this geographic range of diverse habitats wasn't stable over time. Our ancestors' environment underwent several ice ages and pluvial episodes of excessive rainfall, sometimes involving rapid change in a geological sense punctuated by interstadial periods of global warming with changing sea levels. And we don't even need to get into such things as volcanoes, tsunamis, or the occasional meteor strike. These shifting patterns meant alterations in local flora and fauna to which all local species, including humans, needed to adapt to survive. As we know, many didn't make it.

Considering all that, there's a good reason why we're omnivores. Humans happily eat just about anything that won't kill them, and even some things that might. We all have our preferences, and things that we wouldn't eat on a bet. But if those things are edible, somebody somewhere probably does eat them, with gusto. To be sure, there are picky eaters among us, and some who would never dream of serving red wine with fish. But many of us are happy to try new things—cautiously, perhaps, but maybe even just for the hell of it. If we want to talk about our Pleistocene genetic heritage, could it be that this remarkably flexible receptivity to diverse foods could reflect our need in the past to be adaptable? Most other species are far more selective. Just try offering a tuna sandwich to a cow. Beyond doubt, the resilience inherent in this omnivorous capacity extends to flexibility in other aspects of behavior as well.

Does all this, in fact, sound as if the human evolutionary experience would lend itself to fixed, locked-in, behavior modules? That sort of constraint on behavior would have been just as counterproductive for our ancestors as it would be for us. Unless, of course, one would like to argue that some populations have a genetic propensity for eating grasshoppers, others for savoring calf brains, and still others for drinking cows' blood straight from the vein. Yes, we all know about lactose intolerance. But this actually offers a good example of human evolutionary adaptability.

Lactose *in*tolerance, in fact, is the norm for most adult mammals and for the majority of people in the world. The appearance of lactose tolerance in adults, giving some people the ability to digest milk after infancy, is clearly a recent genetic change that arose since the domestication of cattle within the last 10,000 years or so.[32] This evolutionary change in some groups, incidentally, occurred well after the end of the Pleistocene. From what we know of the huge, fierce wild ancestor of domestic cattle, the aurochs, milking such a beast would have been out of the question. One can only imagine trying to milk a wild bison cow or an African cape buffalo.

This development further undermines claims of fixed genetic modules from our far more ancient Pleistocene heritage. Over the past few years, evidence has accumulated to show that numerous adaptive genetic changes have occurred in various human populations over the past several millennia. We didn't finish evolving in the Pleistocene. We're still at it.[33]

It's obvious that the fundamental survival mechanism of human populations in the past, as in the present, has been adaptability—the ability to find a variety of ways to go on surviving. This means passing on a sufficient number of one's genes collectively, as a population, in continually altering circumstances. Thus, for anyone who takes the Pleistocene seriously enough to learn something about it, its impact on human evolution is obvious. Our genome does not specify or determine rigid drives and responses to specific circumstances. It frees us from such impediments. If some early humans ever did have rigid genetic modules for behavior as evolutionary psychologists imagine, they certainly didn't survive to pass on those modules to the present.

GOING OFF SCRIPT

Beyond doubt, our genetic heritage is a result of what happened during the Pleistocene and long before. It includes many ancient features—the "reptilian" part of the brain, for example—that are very primitive indeed. But the human brain also includes an enlarged frontal lobe whose primary function is taking in, storing, and processing information that originates *outside* the skull—in the environment, including the social environment populated with other humans—rather than from within neural modules. It includes what we like to call "knowledge."

The Pleistocene genetic heritage also lives on in the parts of the brain that process language—not only allowing us to understand the utterances of other humans, but to create our own. Although many of the statements we come up with on a daily basis without thinking much about it have never been spoken before, anyone who speaks the same language can understand them without trying very hard—unless, of course, they happen to be the convoluted pronouncements of certain professors, jurists, or other special cases. These are not automatic, reflexive sounds such as barking, squawking, or whimpering, although we can do those too. They're manifestations of a unique evolved ability to transcend biological scripting—the ability to adapt and to do so collectively by sharing information.

That brings up another aspect of sociobiology and evolutionary psychology: the fundamental premise that what's going on in human life at the most fundamental level is competition among individuals who are striving to pass on as many of their genes as possible at the expense of others. Their model of evolutionary survival is starkly individualistic, to the extent that the continuance of the population is almost incidental, an epiphenomenon, a by-product of genetic competition among the individuals within it. Along the way, of course, the fittest supposedly leave more of their genes in the pool than the others who tried but . . . . well . . . didn't quite make it. Their genes weren't invited to the next generation.

Oh, sure—evolutionary psychologists do recognize instances of cooperation. But they account for this almost dismissively by seeing such interactions in terms of "inclusive fitness." They account for what might seem to be altruistic behavior on that basis: sort of "I'll help you, even at risk to myself, because you have a lot of the same genes I do." As we've noted, they also allow for "reciprocal altruism": "I'll help you, but only if I can be pretty sure you'll help me later." This leads to the assertion that, over the course of evolution, we humans have developed genes for "detecting cheaters"—genetic antennae for spotting those who are happy enough to receive help but can't be relied on to come through and reciprocate. Our ability to detect such cheaters, apparently, is the result of a genetic module that also has been selected for.

And all this time, a lot of us were under the delusion that such insights usually come the hard way—through experience. Remember the old saying, "once burned, twice shy"? If we have these genetic detectors, why should we have to get burned in the first place? And we won't even get into the warnings and teachings passed down by parents, wise old aunts and uncles, and grandparents. As long as they've passed down their genes, who needs all that lore and wisdom? With cheater-detecting modules to take care of that sort of thing, who needs the corny old voice of experience? But wait a second. If we have genes to detect "cheaters," shouldn't evolution have worked against cheaters a long time ago and kept them from being able to pass on so many of their low-down, sneaky genes? There still seem to be a lot of them around.

Let's leave the theater of imaginary gene-maximizing robots for a moment and consider some other things we do know—not just about the Pleistocene, but about actual, living human beings.

BATTLE OF THE SEXES?

The mysteries of gender relations have long preoccupied poets, songwriters, novelists, and soap opera script writers. We have numerous literary models to ponder, ranging from poor little Echo's unrequited desire for the self-absorbed Narcissus, or the tragic love of Romeo and Juliet, to that famously dysfunctional couple, the Macbeths. But according to sociobiologists and evolutionary psychologists, the whole thing's not so mysterious after all. Want to guess why not? Hint: It has to do with passing on genes.

This, we can only suppose, is an example of those simple principles with great explanatory power that Pinker refers to as "elegant" and "beautiful." Aesthetics aside, however, it might be useful once again to hold these templates up to the data and see how they match.

In the last chapter, we discussed evolutionary psychologists' use of the insights of Robert Trivers[34] to draw numerous conclusions about aspects of male/female relations arising from the biology of human reproduction. Remember the profligate males and coy, selective females? We might also recall a major study that called this disparity into question,[35] finding that many American women, at least, seemed to have at least as many sexual partners as men. The main difference, it seemed, was that women tended to be more discreet about it. A recent article in the *New York Times* on hook-up culture among college women who willingly discussed the issue suggests that even the discretion part might be so late twentieth century (no longer selected for, perhaps?).[36] So far, evolutionary psychologists seem to be at a loss when it comes to homosexuality for either sex,[37] which turns out to be much more common than many Americans would have thought just a few years ago.

The evolutionary psychology model is far too simplistic on the face of it, since it portrays relationships and associated roles as being far more uniform than what we encounter in real life. It also seems especially ironic that, as Susan McKinnon points out, many of the conclusions that evolutionary psychologists extend to humanity in general are based on data obtained from American undergraduate university students. These specimens of humanity are easily available to researchers, and generally they're willing subjects, but they're certainly not fully representative of humankind in its panoramic diversity.[38] Yet even among that restricted sample, as we've seen in the sexual activity study, the results can be misleading. When we consider the range of the human experience cross-culturally far beyond the campus quad, all bets are off.

David Buss, a leading evolutionary psychologist at the University of Texas at Austin, cites a study that purports to reveal an inherent difference between males and females in the interpretation of social interactions.[39] According to the report, when researchers asked males at a social gathering in a bar near campus about their perceptions of female sexual interest in them, they consistently overestimated the women's feelings of attraction. The women's responses, sadly, did not substantiate the men's optimistic view of things. "The subjects were 144 white Northwest University undergraduates who received credit toward a course requirement of research participation."[40]

Not too surprising, perhaps, though interesting. But is this a finding that we can really assume applies to humans in general, across all cultures? Perhaps it does, but it would take a lot more cross-cultural data to resolve that question one way or another. And that's the point.

Many sociobiology and evolutionary psychology models also entail that women require stable relationships because males control access to resources.[41] Hence, the prototypical woman relies on her mate to provide for her and her children. Okay, sounds like a workable arrangement so far. Perhaps, through looking at various societies, we can get a more detailed sense of just how great this male contribution is to the collective pantry. Hmm.

According to numerous studies of hunting and gathering peoples, it seems that when it comes to food, women's roles are generally at least as important as men's. Richard B. Lee estimates that among the !Kung San, who have been among the examples most cited by sociobiologists,[42] women contributed more than 70 percent of the food supply. Not only that, but their contributions were especially important because they were far more reliable than men's. True, men did most of the hunting, but that enterprise was far less dependable, and the results much more sporadic. It was women who provided food for the family day after day—be it roots they'd dug, berries or other fruit they'd picked, ostrich eggs they'd found, nuts, seeds, or whatever.

What about being tied down with kids at home? It turns out that little ones who were still nursing could come along with mom in the search for food, riding on her hip and sitting in a cross-shoulder bag—often within easy reach of a breast. Others could stay back at camp with grandma and other adults.

The occasional hunk of meat was welcome, of course—but good luck relying on that. And if a large animal did fall to the hunters, they quickly distributed the meat through the multifamily group—or more accurately, through what amounted to a large extended family. (As opposed to the isolated nuclear family that looms large in the evolutionary psychologist imagination, and which, even in the United States, has become common only in the past few generations.) The hunter who had actually killed the animal might end up with the least meat of all.

We find similar patterns among hunters in the Arctic. There, Inuit hunters had numerous elaborate ways of ensuring the maximal distribution of meat, including meat-sharing partnerships with other hunters and, consequently, with their families. Although the Arctic environment had almost no plant foods for women to gather, their work in preparing hides and meat and other vital tasks was crucial for men to survive.

This brings up another serious flaw in the evolutionary psychology model—if we're inclined to consider dissociation with reality a flaw, notwithstanding its "elegant simplicity." Evolutionary psychologists' focus on the family as a pair of mated individuals and their offspring portrays this conjugal unit almost as an analytical isolate. They show little recognition that all people everywhere live in groups larger than the simple nuclear family and always have.

Human groups need some sort of organizing principles and shared understandings in order to go on existing. They need to establish expectations about access to resources and their distribution, methods of conflict resolution, and so on. Yet, in general, sociobiologists and evolutionary psychologists seem to visualize elementary mated pairs existing in a competitive, even hostile, social environment in which the "society" around them consists of similar mated pairs who compete not only for resources, but—wait for it—to pass on more of their own genes at the expense of others.

In that scenario, the most likely reason for such an imaginary society to cooperate or to act collectively would be to attack some other group, or defend itself against one. This is a pretty dismal view of the human experience. The good news is that the model reveals an astonishing ignorance of what life actually is and has been like in most functioning human societies and is not particularly accurate. Certainly, interpersonal and intergroup conflict has always been a part of human history. But a large part of the human experience has also been to develop ways of preventing, limiting, and controlling such conflict. That way, almost everybody gets to pass on more genes.

We might think that before writing broad, universalistic statements about humans in general, one would try to learn something about the subject in a variety of cases rather than relying on a caricature based on a restricted sector of the United States at the beginning of the twenty-first century. But the idea that cultural differences can have profound effects on humans would contradict the evolutionary psychologists' fundamental tenet—that basically it's all genetic, with ancient roots, and that cultural differences are superficial. That leads easily to the assumption that odd customs elsewhere don't really count and aren't worth much attention.

A WORD ABOUT ETHNOGRAPHY

But to return to gender roles and relationships: the model of the hunting, grunting male protecting his mate and their wee babes might be good material for a cartoon, but it has little to tell us about the ethnographic record. And here we might stress the term *record*, meaning accounts of actual observations.

Some might be quick to point out that many of the observations of researchers attempting to learn about other peoples' ways of life in unfamiliar social settings can be subject to misinterpretation, inadequate data, and so on. Too true. But despite the inevitable imprecision inherent in the process of people trying to understand other people, good ethnography involves trying to get it right by actually living within a community over an extended period of time, learning the language and conversing with people at length, sharing their food and their stories, and trying to develop an understanding of what their lives are like and why. And there has been lots of good ethnography. Often, researchers have returned again and again over the course of many years to recheck and update their information. Often, too, other researchers have spent time in the same communities, providing a basis for comparison.

This is not to say that all ethnographies have been equally good or that there haven't been some bad ones. But many have been excellent, and some have become valuable historical resources for later generations of the people whose ancestors they've depicted. They've amounted to gathering information the hard, old-fashioned way. Even if sometimes flawed, they've been infinitely better sources of information than imaginative assertions of people creating visions of "human nature" in their own minds.

What can the ethnographic record show us about gender relations? For one thing, it appears that the isolated couple consisting of one male and one female is only one possible variety, and an extremely rare one, if we visualize this (happy?) couple as a distinct unit surrounded by highly competitive, similar conjugal units. The image of a population consisting of mutually suspicious, potentially hostile small family units evokes the eighteenth-century image of Thomas Hobbes's "war of all against all."[43] It shouldn't be surprising that Steven Pinker argues explicitly in favor of this Hobbesian model in opposition to the ideas of social contract theorists such as John Locke or Jean-Jacques Rousseau, with their concepts of social reciprocity and rational, mutually supportive interactions.[44]

Not only are people in all societies immersed in a sea of social relationships, but generally these relationships function to provide mutual support and resource distribution. Nor is the family unit necessarily male centered. The ethnographic record offers a diverse array of cases with almost every imaginable sort of arrangement, many of them with women at the pivotal points of organization. Some might consider all of these cases to be inconsequential anomalies, aberrations, or departures from "true" human nature. Alternatively, we could see them as multiple examples of the vast range of patterns that our adaptable human nature, freed of the evolutionary psychologists' panoply of rigid genetic constraints, has produced.

We can see, then, that the reverse engineering of evolutionary psychologists has its shortcomings.[45] We recall that the process involves taking a phenomenon that characterizes present human life in a very specific locality and, on the assumption that almost every trait has a genetic basis, then projecting it back to imagine what sort of factors during the Pleistocene "must" have "selected for it." We've seen some of the methodological problems with this.

There's no real evidence that genes dictate any behavioral trait. Evolutionary psychologists' version of the Pleistocene is far too simplistic, and as a consequence, misleading at best. And as it turns out, even their depictions of the present are oversimplified and misleading to the extent that they purport to depict pan-human behavioral traits, since they ignore a vast panorama of alternative cultural patterns.

Their message, of course, is not to celebrate or even acknowledge the array of human diversity and adaptability, but to portray a fairly stark uniformity. In such a world, what we see around us is what has to be. The target audience is not those who hope for change, but those who take comfort in what they suppose has always been and, they hope, will continue to be.

The evolutionary psychology model—the male breadwinner, the stay-at-home mom, and so on—to the extent that it has any validity even for American culture, is fairly recent even in our own history. Ironically, it seems to be increasingly passé even now, as more women become the major breadwinners in the family and more dads stay home to take care of the house and kids. And as for the past, one can only imagine the reaction of a woman working with her husband to run a family farm—which was the typical way of life a couple of generations ago—upon being informed that she's relying on her mate to provide her with resources. In short, they've gotten the present wrong; they've gotten the Pleistocene wrong; and the rest is pretty much fantasy.

Chapter 8
It's Not That Simple

OUR URGE TO EXPLAIN OURSELVES, TO UNDERSTAND OUR lot in life, our wishes, hopes, drives, and experiences and those of our neighbors, colleagues, friends, and enemies, probably has always been with us as a "sapient" species. We've spent a good deal of time here showing that persistent attempts to reduce this question to biology have not only been inadequate; many of them have done considerable harm.

As Hilary and Steven Rose note, recent versions of biological determinism in the past few years have re-essentialized sex, gender, and "racial" differences, just at a time when many of us hoped we had begun to move beyond that era.[1] It might have been tempting to begin this chapter with a statement about having completed a long journey over the history of attempts to impose biological explanations on human life. But as with the exhausted little kid in the back seat of the car, the more appropriate question seems to be, "Are we there yet?"

The past several decades have been a crucial period in the continuing development of American society. Title IX laws in the United States have opened an entire new realm of athletic achievement for young women in schools, after generations in which the conventional wisdom was that girls are too delicate to engage in major sports or that women are naturally less competitive than men. The open sexism of the 1960s has now become anachronistic material for entertainment on television. Civil rights struggles in recent decades have made the Jim Crow era seem a horrifying chapter of ancient history. But this certainly doesn't mean we've reached a safe, comfortable plateau. This is also a time when millions of women and ethnic minorities continue to struggle for equal treatment in the workplace, including fair wages and career opportunities. It's a time when the election and re-election of the first president with African ancestry, which many viewed as a triumph over our own history, has given rise to an upsurge in militant racist rhetoric and organized hate groups.[2] All of this has shown us, in the words of William Faulkner, that "the past isn't dead; it isn't even past."[3]

Clearly, this is a time in our society, like many others before, when the direction of the future is a contested issue. It's a time when evolutionary psychologists reassert that profound, evolved individual differences in ability, strategic decision making, and aggression are an innate part of a biologically transmitted script—the directions that come with the equipment. Such a view doesn't allow much room for optimism about change. Apparently, we are the way we are because that's the way we are. But it may be useful, on that note, to recall that in many important respects, we're no longer the way we used to be. We've already changed quite a bit in a short period of time.

Over the past century or so, depictions of human nature have resonated with the state of biological science of the time. Attempts in the early twentieth century to visualize different cultural and ethnic populations or social classes as different subspecies of humanity now appear astonishingly crude and ignorant. More recent arguments have employed the language of genetics, embellished with a terminology crafted for the occasion and apparently intended to sound dauntingly scientific. In many respects, contemporary versions have consisted more of updated jargon and poses of scientific rigor than of the actual incorporation of current knowledge.

This is particularly apparent in evolutionary psychology, where multiple assertions of a gene for this or that highly specific form of behavior—or as Hillary and Steven Rose put it, "daily claims to have identified a gene for almost everything"[4]—have become more and more implausible, perhaps even indefensible, with rapid advances in the science of genetics. Although the vocabulary is intended to convey an aura of complex and highly abstract "hard science," genuine science itself has led us away from these sorts of deterministic assumptions.

We do come across an interesting back-and-forth on the issue of homosexuality. Evolutionary psychologists don't seem to have had much luck in coming up with a plausible explanation for why such a characteristic would have been selected for in the Pleistocene. Yet by the second decade of the twenty-first century, as states repealed numerous barriers to the legal recognition and access to equal rights for homosexual individuals and couples, it turned out that homosexuality had been a far more common part of the human condition than many would have imagined. In that context, a debate has raged over whether homosexuality is a matter of choice or an inborn characteristic.

In an apparent reversal of roles, many conservatives have insisted that sexual preference is a matter of choice—of people choosing that particular "lifestyle." There have been movements to "pray away the gay" and other efforts to help people overcome this "dread affliction." Most recent scientific opinion, however, is that this characteristic is inborn. Many liberals have had no problem in accepting this idea.

In response, some biological determinist conservatives have seen this as a "cave" on liberals' part, as if an admission that any trait is inborn amounts to accepting biological determinism in its entirety. But no one, of course, would argue that no human characteristics at all are inherited. That would be ridiculous. The issue is the extent. Perhaps the greater irony is that biological determinists, in this case, would greatly prefer that homosexuality is a matter of choice, a willful departure from the "the way it ought to be," because one could then argue that being voluntary, it might warrant sanctions or punishment.

This also ties in with the old "family values" argument. Many conservatives seem to cherish the image of the old-fashioned, "natural," one man–one woman ideal. This view opposes change and embraces the status quo or, in many cases, a fanciful status quo ante that never really existed. That's without taking into account that in actual human history, cross-culturally, the monogamous nuclear family has been only one of many varieties of family structures that include polygyny, polyandry, and other interesting arrangements.

Recent achievements in human genetics have underscored the complexity of gene interactions and epigenetic factors involved even in straightforward physical, nonbehavioral phenotypic features. As the Roses point out,[5] well over a hundred genes may be involved in just one particular aspect of blood chemistry. As to behavioral issues, they note that more than 150 genes seem to be involved in schizophrenia, which is, after all, a pathological condition. How many more would have to be involved in aspects of what we might think of as complex patterns of "normal" behavior?

This sort of insight isn't entirely new. Working with fruit flies in the 1920s, Herbert Jennings noted that "at least fifty genes are known to work together to produce so simple a feature as the red color of the eye; hundreds are required to produce a normal straight wing, and so of all other characteristics."[6] Jennings insisted that this complexity was something that eugenicists should keep in mind.[7]

The Human Genome Project has shown that we have far fewer genes than most researchers would have assumed just a few years ago probably between twenty thousand and thirty thousand, as opposed to one hundred thousand or more as many geneticists once estimated. Only a few of these, from 1.6 percent to 4 percent, differentiate us from chimpanzees.[8] These genes—our own special package—number only in the hundreds, not in the tens of thousands. If the evolutionary psychologists were right, that would be a lot of work for just a few genes to accomplish. It's clear that genes work in multiple combinations and with a range of epigenetic interactions. As a crude analogy, there are only twenty-six letters in the English alphabet. It's the nearly endless combinations of these letters that allow us to write far more than twenty-six different things. Given these parameters, to suggest that there's a gene (or genetic module) for sharing, for sexual jealousy, for male aggression, for worrying, for just about *any* complex behavior pattern sounds even sillier now than it ever did.

As one of the main researchers in the Human Genome Project has expressed it, "We've called the human genome the blueprint, the Holy Grail, all sorts of things. It's a parts list. If I gave you the parts list for the Boeing 777, and it has 100,000 parts, I don't think you could screw it together, and you certainly wouldn't understand why it flew."[9]

We've speculated about reasons for the persistence of the idea of a biological script over the past century and a half even though it's been discredited in every era by the science of the day. Biological determinism does provide simplicity in accounting for complex forms of behavior, which some might view as appealing. But it also implies, or even asserts, an inevitability of various types of behavior that some of us might hope to change: gender inequity, aggressive and selfish behavior, wealth inequality, racism, sexism, and other discriminatory patterns.

SO WHAT'S THE ALTERNATIVE?

Having discussed the shortcomings of biological explanations, what can we offer instead? Not too much, as it turns out. No one needs to be told that human behavior is complicated or that it occurs as a result of myriad factors. That's just the way we work. Why things happen, why people do what they do . . . it all depends. To put it a bit too simply, perhaps, people are not that different from one another genetically, but they do lots and lots of different things. Even the same person often behaves differently at various times. Whatever makes us act in these ways, the reasons are a lot more complicated than marching orders from our DNA.

We can all think of examples of entire nations apparently changing their characteristics from one generation to the next. A country that presented a bellicose posture and threatened world peace in one era might be a peaceful global trading partner in another. Surely, this can't be due to genetic change. Even the same population, separated by political boundaries and different systems of governance, can display contrasting characteristics. And whatever happened to the Vikings who terrorized their part of the world just a millennium ago, sailing out from the region where Swedes and Norwegians now enjoy peaceful, progressive societies? Obviously, the genetic makeup of these populations hasn't changed drastically in just a few generations.

Evolutionary psychologists, after all, maintain that we're still ruled by paleo-genes that were selected for back in the Pleistocene. But if we did have genes or modules directing our behavior, they certainly couldn't be molding or determining behavior in any rigid way, since they haven't prevented radical changes in behavior from occurring over and over again. And so we might wonder, Why would it be necessary even to imagine they exist?

Those examples of collective behavior focus on a fairly macro level. It's obvious that many factors have played a role in changing the apparent characteristics of different populations. Economic relationships, wars, political arrangements, resource issues, and many other elements have combined to bring about these outcomes. Few of them, if any, have to do with genetics.

At a more micro level, evolutionary psychologists also have much to tell us about the behavior of individuals. But as we've seen, these findings don't allow us to predict the activities of particular persons on the ground, in real time, with any reliability. Some males dominate their wives. Some wives dominate their husbands. Many couples have relatively congenial complementary relationships, not to mention the many societies in the world in which the monogamous nuclear family isn't even the standard model.

Some aggressive males may have had abusive childhoods or been spoiled as children, or be drunks or drug abusers, or have grown up hanging around with the wrong crowd or seen the wrong kinds of movies, or, for whatever reasons, just be nasty bastards. But lots of males are not that way. Countless elements and factors go into making people who they are and the way they are, how they develop, and even how they change throughout their lives. Most of them have little to do with genes, except that our genes make us creatures who are highly receptive to learning from, adapting to, and being influenced by our experiences.

AH, TRADITION

Histories of ideas can be instructive. In the debates for and against biological determinism, we can see at least two major opposing traditions. One, embraced by Steven Pinker and others before him, has roots in the writings of Thomas Hobbes. Pinker writes favorably of Hobbes in opposition to those he refers to as "blank slate" foundational thinkers such as John Locke and Jean-Jacques Rousseau.[10] Hobbes saw humans as brutish and self-interested by nature, requiring the imposition of firm constraints for society to function. Hobbes's concept of the "leviathan" depicted an overarching administrative power that kept the "war of all against all" under control. If we jump forward in history to Herbert Spencer over a century later, we can see a hint of the leviathan in his "organic analogy" of society.

In Spencer's version, societies as organisms follow natural laws. Natural laws not only constrain individuals, who live in competition with one another. Societies also compete with other societies for survival. This reign of competition from the cellular level upward is the driving force of progress, in the course of which some individuals and some societies fail to survive. The fittest among them not only continue to exist, but thrive—a fact, that, in itself, attests to their superior fitness. Some societies even attain greater complexity, progressing from small undifferentiated bands to state systems. Natural law, in Spencer's view, generates progress through competition. Spencer gave us social Darwinism, but the roots of these ideas drew on the previous century, from ideas that Hobbes expressed to such great effect.

The heritage of Spencer in sociobiology is clear. Like Spencer, E. O. Wilson envisioned an overall science based on biological principles which, when fully realized, would account for just about every aspect of human life. Eventually, it would subsume the social sciences and humanities within the rubric of what Spencer would have referred to as natural law. And like Spencer, Wilson saw competition as the driving essence of evolution, and hence the underlying principle of life itself. Unlike Spencer, however, Wilson paid little attention to society and focused primarily on the individual organism.

Wilson had the advantage of knowing about genes—an insight that wasn't available to Spencer. For Wilson, individuals, as the carriers of genes, were the main active competitors in the arena of life, but from an evolutionary perspective, the point of individuals' existence was to achieve success in perpetuating the genes they (unknowingly) carried. Society (and culture), in Wilson's view, were essentially epiphenomena in the life process. Most of the characteristics of life in a given time or place were the results of genes having been selected for through competition at the individual level. Not to worry, however. The fittest would survive.

Eventually, the somewhat overextended assertions of sociobiology fell out of favor, and evolutionary psychology took up the cause. As Susan McKinnon and Sydell Silverman note,[11] while sociobiologists tended to see humans as just another species and to generalize from animal to human behavior and then back again, evolutionary psychologists generalize from the behavior of a few humans to an imaginary Pleistocene era and then back again to humankind in general, assuming these examples of behavior to be human universals. The crucial role of genes in scripting behavior remains in the forefront, however.

We can see a different and opposing tradition rooted in the ideas of Locke and Rousseau. Pinker explicitly rejects these thinkers in favor of Hobbes, holding them largely responsible for the pernicious concept of the blank slate. The blank slate, as we've noted, was a metaphor to emphasize the fundamental equivalence of all humans and attributes differences among them to the diverse processes of learning and life experiences.

As we follow the trail through nineteenth- and early-twentieth-century thinkers, one of the more interesting was the Russian anarchist Prince Peter Kropotkin. Like Herbert Spencer, Kropotkin drew inspiration from Darwin, but he took a very different lesson from Darwin's insights. While Spencer had focused on competition at all levels as the central theme, Kropotkin concluded that in the struggle for survival, those groups who were best able to cooperate among themselves would have an advantage over others.[12]

Kropotkin never achieved the celebrity that Spencer enjoyed, although he did participate in debates with Spencer, and at one point, he was an honored speaker at the Royal Geographical Society. Perhaps Kropokin's failure to achieve greater renown was partly because the term "anarchy" had less popular appeal than unregulated capitalist enterprise, for which Spencer's ideas are more congenial. Perhaps, too, it had to do with others who also called themselves anarchists at the time who had a reputation for throwing bombs in crowded cities.

Unlike the bomb-throwers, though, Kropotkin held a position that was far more peaceful. He opposed government because he believed that such power structures interfere with the natural tendency of humans to live in harmony and mutual assistance. Such ideas, of course, stood in stark contrast to the hypercompetitive models of Hobbes and Spencer, even though both Spencer and Kropotkin found inspiration in Darwin's work.

Despite Kropotkin's relatively low profile in history, his ideas flowed into the intellectual milieu of the early twentieth century. He had at least some indirect influence on the British anthropologist Bronisław Malinowski, who refers to him in passing,[13] and very probably on the man many consider the founder of British social anthropology, A. R. Radcliffe-Brown.[14] They, like most anthropologists at the time, focused their attention on smaller nonstate societies and their capacity to function without central government. Many of these groups did, in fact, exist under European colonial rule in that era, but most anthropologists attempted to understand how such societies could function, or had functioned in the recent past, without the imposition of outside governance. They attempted, in other words, to begin to understand these societies as functioning entities in their own right, rather than as obstacles to progress, as many others saw them at the time.

In the United States during the same period, the concept of culture predominated in anthropological thinking. In both cases, the emphasis was on people's capacity to learn, to structure, and to organize their lives in a variety of ways without any assumption of biologically differentiating factors. Although Franz Boas did help establish physical anthropology as a subfield of anthropology, explicit in all of these approaches was a rejection of biological explanations of behavior and of racist explanations in particular. For most of the nineteenth and twentieth centuries, as we've discussed at length, the latter two have often been entangled.

SOMEHOW IT ALL FITS

We've seen some of the ways in which these competing ideas have come to prominence at certain times and faded in others. Spencer was an intellectual celebrity of sorts in an era of laissez-faire economics and colonialism, an ideological climate that celebrated

Darwinian competition in the marketplace. Kropotkin's rather muffled alternative views stressed cooperation. These may have caused alarm in a few quarters, perhaps seeming to smack of socialism,[15] but they remained a part of the background in the intellectual milieu of the time.

The orientation of Kropotkin, Radcliffe-Brown, Malinowski, Boas, and Kroeber only came to the forefront in the first few decades of the twentieth century. In the 1930s, the Roosevelt administration began to enact social programs to overcome the catastrophic aftermath of the excesses of the Spencerian "Gilded Age," an era of raw social Darwinism. Public projects to build roads, dams, parks, and public structures provided jobs for tens of thousands of the unemployed. Social Security helped to protect retired workers from starvation in their old age.

One could see this development as the workings of a sort of benevolent "leviathan" (though it seems highly doubtful that any of its proponents would have referred to it that way), with the structure of governance assisting the populace, but the emphasis was more on meeting human needs than on enforcing peace among a belligerent Hobbesian competitive rabble. In another sense, however, many proponents of the New Deal did see these measures as a means of forestalling serious social unrest.

From that era through the 1960s in the social sciences, an approach based on the analysis of human learning, choice, strategies, and decision making predominated. Hobbesian views popped up occasionally in, for example, Robert Ardrey's "killer ape" model and Konrad Lorenz's arguments for innate aggression. They provoked some discussion in their day, and as we've seen, drew some positive media coverage. But in the academic sphere, they remained marginal. In biology, evolutionary studies at the time still focused on populations rather than on individuals, and psychology still had a behavioral, experiential focus. After World War II, increasing awareness of the Holocaust had discredited overt expressions of racism and calls for eugenics programs in most quarters. This was the era that had opened with Franklin D. Roosevelt's New Deal and continued through Lyndon B. Johnson's Great Society programs of the 1960s. That phase of American history saw many intense and sometimes bloody conflicts, including the Red Scare of

McCarthyism and struggles for civil rights. But an overall trend was away from the deterministic thinking of the past.

By the 1970s, however, the ideological climate had begun to change again. Richard Nixon was in the White House. The Cold War, with its chronic vague but looming cloud of anxiety over the possibility of mass destruction, was about to enter its third decade. The Vietnam War continued to drag on, with tens of thousands of American dead and no end in sight, despite nightly enemy "body counts" and the long, now clearly failed, campaign to win Vietnamese "hearts and minds."

A major oil crisis resulted in the widespread loss of jobs. Many directed their anger at the OPEC oil cartel and at foreign auto imports, both of which seemed to symbolize outside interference with American well-being. The Vietnam War had deeply divided the country, as did continuing civil rights struggles, including mandatory school busing to promote racial integration. Angry crowds in South Boston threatened children on school buses.

It was a time of looking for causes and explanations and directing blame. In many ways, it also involved a repudiation of the social and political initiatives of the 1960s. A long campaign by conservatives in Congress had begun to roll back many of the social programs of earlier years. It was a time ripe for scapegoating and "stern measures," with plenty of "law and order" candidates running for political office.

Wilson's *Sociobiology* appeared in 1975, with a social message resonant in many ways with the tradition of Hobbes and Spencer, emphasizing competition among individuals as the mechanism of evolutionary progress.

In much popular thinking over succeeding decades, some on the political right came to consider the overall structure of "big government," which had provided social programs from the 1930s through the 1960s, to amount to an exercise in illegitimate social engineering. Big government, rather than helping to keep your grandmother alive, now threatened to "take away your freedoms."

As the leviathan evolved from Hobbes's stern enforcer of discipline to become FDR's benevolent caretaker, it became a monster under the bed for extreme libertarian conservatives who resented anyone telling them what to do, whether it be paying taxes or having to serve people with different skin color at their lunch counters. This was particularly so following the passage of civil rights legislation in the 1960s.

From the 1980s onward, in an era when Congress had begun to make severe cuts in funding for social programs, we saw sociobiology evolve into evolutionary psychology, which in turn paid little attention to collective phenomena and stressed individual behavior based on innate, fixed drives.

By the first decade of the twenty-first century, the country had become yet more divided—still, as in the past, largely on a regional basis. Although Barack Obama won the presidency twice with clear majorities, opposition on the far right became still more entrenched. Congress stripped away numerous government regulations that had been enacted in previous decades, wealth disparities increased dramatically, and some saw a perilous trend back toward the laissez-faire imbalances of the previous century's Gilded Age. In 2013, the House of Representatives voted to reduce the budget for food assistance to the needy by almost $40 billion, at a time when abject poverty was increasing dramatically. A month or so later they cut off federal funding altogether by refusing to pass the federal budget.

As these examples show, the Spencerian theme of progress through competition saw a resurgence that continued well into the twenty-first century, into an era when few people any longer had the slightest idea who Spencer was. But as historian Richard Hofstadter wrote in the 1940s, "If Spencer's abiding impact on American thought seems impalpable to later generations, it is perhaps only because it has been so thoroughly absorbed."[16]

The contemporary icon for these ideas became Ayn Rand, a mid-twentieth-century author whose novel *Atlas Shrugged*[17] became a sacred text for many of those opposed to social programs and big government. Her co-authored collection of essays had a slightly more informative title: *The Virtue of Selfishness*.[18]

Conservatives such as Wisconsin Senator Paul Ryan, vice presidential candidate in 2012, referred to Rand's work as "required reading" for his followers. Herbert Spencer in the nineteenth century had argued against collectively funded institutions such as public libraries and schools on the grounds that they interfered with the natural process of elimination of the weak. Herbert Spencer put it rather bluntly: "The whole effort of nature is to get rid of such, to clear the world of them, and make room for better."[19] As Richard Hofstadter notes, "Spencer not only deplored poor laws, but also state-supported education, sanitary supervision other than the suppression of nuisances, regulation of housing conditions . . . tariffs, state banking, and government postal systems."[20]

The Rand-inspired twenty-first-century conservatives fought against government social programs ranging from medical care to food aid for the needy. Many also sought to "privatize" the US Postal Service. The "Ryan budget," a plan to withdraw funding from a range of federal social programs, had won nearly unanimous approval of the Republicans in the House of Representatives in 2010. By 2013, the intransigence of conservatives in Congress in opposing social assistance programs—and their animosity toward twice-elected President Obama—ultimately brought about the temporary shutdown of the federal government.

More and more, biological determinism crept back into the public arena. In 2013, the conservative Heritage Foundation faced public embarrassment when it commissioned a report on immigration policy, a topic that at the time was under intense national discussion. Not surprisingly, perhaps, the report enumerated a range of objections to more moderate policies for Latino immigrants. One of the co-authors, Jason Richwine, had expressed some explicitly negative views on Latinos in previous publications. Many of Richwine's arguments focused on the old issue of IQ. He noted that not only were the scores of Latinos lower than those of other groups, but claimed they would remain so for several generations at least because of the degree to which IQ is inherited. His PhD dissertation at the Harvard Kennedy School had addressed these issues at length. Among his numerous citations, a few familiar names jump out: Richard J. Herrnstein, Arthur Jensen, Charles Murray (nine citations), and J. Philippe Rushton.[21]

It might seem discouraging in some quarters that such threadbare arguments have continued to appear, even in supposedly reputable contexts. But perhaps we can find some encouragement in the wide-spread outrage that this news provoked. The Heritage Foundation, in the process of enhancing its already ultraconservative image under the leadership of the extreme-right former Senator Tom DeMint, nonetheless found it expedient to fire the embarrassing Dr. Richwine. The Republican Party was at the time, after all, trying to develop strategies to win more Latino votes.

Chapter 9

WHAT'S THE AGENDA?

IT WOULDN'T BE TOO FAR OFF THE MARK TO SEE MANY OF these reactionary initiatives as manifestations of a new eugenics. The primary victims of the gutting of social assistance programs have always been, of course, the most disadvantaged of society—those most in need of food subsidies, medical care, and educational support. Republicans in Congress, while voting to increase subsidies for large agricultural industries in 2013, had voted fifty times to repeal the Affordable Care Act and cut $40 billion from funding for food subsidies for needy families. In the course of budget cutbacks, Congress not only cut off funding for Head Start, a program to assist disadvantaged preschool children that we discussed in chapter 4, but Meals on Wheels, a program that brings hot meals to homebound elderly people.

Republicans also insisted that welfare recipients submit to drug tests—in a year in which several prominent members of Congress and other highly placed politicians were caught using illegal drugs. Many of those affected by the cuts in social programs were Hispanic and African American, though the majority of them were not. Those most affected were people representing a variety of backgrounds. What they had in common, of course, was that they were the most vulnerable members of society, generally because of poverty or disability.

Many in the country gaped in amazement at what appeared to be a festival of mean-spiritedness. For many, the motivation seemed unclear. The costs of these programs, after all, were far smaller than the largesse Congress has bestowed on major corporations and the military on a regular basis, even to the point of insisting on paying for weapons systems that the Pentagon had repeatedly stated it didn't want. No doubt the manufacturers of these systems in congressional home districts were happy to see the expenditures continue.

Early in 2013, after losing the presidential election for the second time, the Republican Party produced a study they referred to as the "post mortem," which laid out strategies for success in the future. Among the recommendations were greater efforts to win the support of African Americans and Latinos, who had voted overwhelmingly for Democrats in the recent election. Yet in the face of such words of wisdom from the party leadership, Republican politicians continued to push measures that directly harmed these constituencies and the disadvantaged sections of the population in general. The two trends seemed contradictory. But perhaps closer attention to the rhetoric can help clarify things.

As some observers have noted, not only did Republican-sponsored policies harm the disadvantaged. Many Republicans and their spokespersons also worked strenuously to demonize those who stood to suffer most. One often heard the phrase "takers versus makers." Recipients of food stamps were cast as a lazy class of people "getting free stuff," even though statistics showed clearly that many of them were working two or even three low-paying jobs but still needed supplemental assistance to feed their families. Rhetoric portrayed such people as drains on society, taking resources that "the rest of us had to work for," or as people "living off my taxes." And, of course, the "racial" and ethnic undercurrents of this, not always explicit, were nonetheless very clear.

We can see this contemporary tradition going back to the presidency of Ronald Reagan in the 1980s, who famously conjured up the image of the "welfare queen" driving an expensive car. Reagan began the era of cutbacks in assistance to the poor that by 2013 had begun to resemble a feeding frenzy. But the rhetoric was reminiscent of a still earlier era. Much of it resonated with echoes of the heyday of eugenics early in the century before. At that time, too, a dominant theme had been the parasitic effect of the poor, the newly arrived, the "different" on society.

Among eugenicists at that time, as among right-wing Republicans in 2013 and 2014, there was little or no suggestion of helping these people.

That would only encourage them, possibly even allow them to overrun the country by reproducing even faster. The solution, rather, was to suppress the threatening tide. We recall from chapter 2 how eugenicists attempted to accomplish that goal. Some of their proposed solutions were involuntary sterilization, mass detention, restriction of immigration, and, in some cases, euthanasia. Some of these measures became a terrible reality. As Edwin Black puts it, "Eugenics was nothing more than an alliance between biological racism and mighty American power, position, and wealth against the most vulnerable, the most marginal, and the least empowered in the nation."[1]

In the twenty-first century, many of these issues seem eerily familiar. As early as 2001, the historian of science Garland Allen wrote about the early twentieth-century eugenics movement and went on to note that

I would argue that we are poised at the threshold of a similar period of our own history and are adopting a similar mind frame as our predecessors. A "bottom line" mentality is rapidly becoming our guidepost. It is unlikely that we will see a return to blatant demands for sterilization, but the requirement of antifertilization medication for continued welfare benefits in the U.S., and the bitter anti-immigration sentiment in the southwestern U.S. and Europe are haunting reminders that we are not immune to the prejudices of our predecessors.[2]

In September 2014, former Arizona state senator Russell Pearce recommended that recipients of Medicaid be subject to mandatory sterilization. After his quote that if he were in charge, "the first thing I'd do is get Norplants, birth-control implants, or tubal ligations," he resigned from his position as his party's first vice chair. Pearce's comments caused outrage in some quarters and embarrassment in others.[3] For some, no doubt, his major offense was excessive candor. Some may recall Pearce as a major force in pushing through Arizona's draconian 2010 immigration law, the harshest immigration law in the country.

Illegal immigration in the twenty-first century motivated the construction of massive walls along the southern border. The Border

Patrol expanded to the extent that these "public guardians" increasingly began to harass people going about their daily business in areas that happened to be near the borders (both Mexican and Canadian), setting up roadblocks and questioning everyone passing through—often not only asking ordinary citizens where they were going, but why, and what they going to do there.[4]

SOLUTIONS THAT CAUSE PROBLEMS

By the twenty-first century, the old methods of involuntary sterilization had been discredited as a national embarrassment, even though they continued in California well into the 1970s. Other means of discouraging the growth of these burdensome elements of society remained, however. With high unemployment and increasing poverty affecting society disproportionately and wealth being increasingly redistributed upward, the curtailing of social assistance was bound to have devastating effects on a growing portion of the population. By 2012, in some important vital statistics, the United States had dropped significantly among developed nations.

Life expectancy, though it had increased overall, did so unevenly. "Even as the nation's life expectancy has marched steadily upward, reaching 78.5 years in 2009, a growing body of research shows that those gains are going mostly to those at the upper end of the income ladder."[5] And as every metric shows, over the past few years those at the upper end of the income ladder have become a smaller, more distant group of elites far above the rest of the population. Despite the gradual overall increase in life span, the United States ranks twenty-sixth among developed nations.[6]

The number of women dying in pregnancy and childbirth has risen since the late 1980s. By July 2013, at 24 per 1000 pregnancies, the United States ranked below Greece, Japan, and Great Britain. It was tied with Saudi Arabia. The rate for African American women was 3.2 times the national average.[7]

Infant mortality also rose. The Save the Children Foundation reported in 2013 that "poverty, racism, and stress are likely to be important contributing factors to first-day deaths. . . . And Republican lawmakers have continued to target Planned Parenthood, advancing measures to defund this national women's health organization at the expense of low-income women who rely on those clinics for their primary care."[8]

All of this has occurred at a time when medical science had advanced tremendously. Many cases of infant mortality could have been resolved through proper medical treatment or prevented in the first place by adequate nutrition. Draconian cutbacks in public funding for such purposes, as well as the radical upward redistribution of wealth, forestalled such measures. Wealthier sectors of society generally didn't need to face such problems, which fell most heavily on those most in need of public assistance. It may no longer be acceptable to sterilize people en masse against their will. But there are other ways of reducing their numbers, including just letting them die.

Much of this, of course, has rested on the continued, deep-seated conviction in some quarters that those at the lowest ranks of society are inherently different and basically incapable of change. Helping them would only increase their numbers and worsen their drain on social resources. Society would be better off without them. If they aren't fit enough to survive without outside help, so be it. As Herbert Spencer put it a century and a half ago, "If they are not sufficiently complete to live, they should die, and it is best they should die."[9]

Differential rates of population increase carry a special irony, considering the history of biological determinism. From the early social Darwinists onward, it's been about the fittest passing on more of their genes by having more surviving offspring. Matt Ridley asserts that "wherever you look, from tribal aborigines to Victorian Englishmen, high-status males have had—and mostly still do have—more children than low-status ones."[10]

That would make sense, if sociobiologists and their intellectual heirs are right about the race to pass on genes. But wait a minute. Has anyone actually documented this? "Wherever you look" sounds more like an impression than evidence—especially when the range is wide enough to encompass everyone from "aborigines" to Victorian Englishmen. And weren't many of these same Victorian Englishmen concerned about the likelihood that just the opposite was going on? One of them, Herbert Spencer, took pains to account for it by calling attention to the competition between reproductive cells and brain cells. Another, Francis Galton, initiated the eugenics movement to encourage more "high status" males to get busy and sire more superior children.

The idea of competition to reproduce would suggest that those who can, will. But it seems that it's a little more complicated than that. Sure, we have the intriguing case of Genghis Khan, who apparently sowed one distinctive sex-linked gene so widely that it now shows up in males throughout much of Europe. But strangely enough, nowadays many of the most prosperous ("high status?") countries in Europe are facing a different sort of issue. They're not reproducing enough even to maintain existing population levels.[11]

Perhaps this would be an appropriate time to cue certain nonbiological factors, such as culture and economics, onto the stage. It would seem that relative prosperity and higher levels of education have turned out to be important birth-control factors. In many poorer countries, it's become almost axiomatic that one of the most effective means of controlling overpopulation is to promote the education of girls and career opportunities for women.

THE BEAT GOES ON

And so the historic process continues. What may seem at one level to be an academic argument—an endless squabble among professors with little connection to the "real world"—it is, and always has been, far more than that. It's a contest over the very nature of that real world—the world in which things happen to people and in which social policies, power structures, and expectations can affect, enhance, or ruin lives.

It is no coincidence that an upsurge in assertions of biological determinism coincides with neoliberal economics and the politics that go with it. The term *neoliberal* does not carry the common meaning of "liberal" as synonymous with "progressive" or "egalitarian." It harkens back to an earlier meaning: a lack of regulation in the market, allowing ruthless competition, essentially replicating the laissez-faire economy of the Gilded Age when wealth disparities in the United States were almost as extreme as they've become at present. That era, in the minds of some, was a golden time before labor unions, workplace safety regulations, the forty-hour workweek, or child labor laws. The main difference in the modern era seems to be that major corporations, rather than being left entirely on their own, receive massive government subsidies and tax exemptions.

Razmig Keucheyan links "the rise of neo-liberalism with the election of [Margaret] Thatcher and [Ronald] Reagan" in Britain and the United States. He notes that "the initial oil shock of 1973 heralded a difficult time economically and socially, with the first significant increase in the rate of unemployment" since the 1950s.[12] This energy crisis produced a feeling of victimization among many of the populace, which some politicians found easy to exploit, particularly by identifying scapegoats among the most severely victimized of all.

We've seen a persistent drive to privatize social institutions that had long been a matter of collective responsibility, ranging from public schools and the postal service to Social Security retirement programs. The underlying philosophy has essentially been that if we were to open these institutions to competition in a free market, they'd become more effective and efficient. The fundamental article of faith, of course, is that the fittest will always survive. It remains unclear, however, whether in such a competitive arena, "fitness" means providing the best service at the least cost or merely eliminating competition, perhaps through cost-cutting, profit-taking, and minimizing expenditures. This, of course, would mean reduced services and increased prices. No "social contract" here, in the Hobbesian neoliberal market.

The current phase of biological-determinist "human nature" arguments, as we've seen, stresses competition at the individual level for passing on genes. It also implies that substantive social change, to the extent that this would involve altering human behavior, is unlikely because behavior modules selected for in the Pleistocene have been locked in ever since. This view of humanity is, of course, quite compatible with the hyperindividualism embraced by the extreme political right wing, with its tea party–infused rejection of government.

REFLECTIONS ON THE MYSTIQUE OF SCIENCE

In the ancient human quest to understand ourselves and one another, science in the past few centuries has enjoyed a special mystique. In a sense this seems particularly ironic, given the history of science in helping to dispel mysticism through the use of evidence and observations of the "real world." But in some respects a different sort of mysticism has managed to don the cloak of science, hiding under its cover to convey an aura of scientific validity, lending credence to its pronouncements of social philosophy.

The idea that the imprimatur of science confers authority on social pronouncements, rendering them immune to questioning, involves a deep irony. If ever the term *survival of the fittest* applied anywhere, it applies to genuine science. Science develops through questions. It depends on testing ideas by trying to disprove them. This process can be long and tedious, but it has little to do with making broad assertions about human nature based on meager data from small nonrepresentative samples or, worse yet, from what "seems logical" based on what we imagine happened during the Pleistocene.

As we know, the genuine answers as to why people do what they do are difficult to come by. They're elusive, complicated, incomplete, and often unsatisfying. What a relief it must be to leave it to "experts" who've been working on the problem and can offer some clear, dispassionate answers.

And since these answers invoke genetics, one of the most complex sciences of all, they not only carry weight; perhaps they can even offer comfort. This is the heart of the conservative thrust of the biological determinist position. It's not just that certain specific changes might not be desirable or feasible for some particular reasons—whether those reasons are political, economic, moral, or whatever. More than that, it rules out change itself. Significant change just can't happen, thanks to that Pleistocene selection stuff. Human nature became what it is a long, long time ago, and we didn't get a vote. So what we do or fail to do isn't really our fault. It's what happened back in the Pleistocene that makes us act this way. We're just burdened with leftover genes from a crueler, more savage time.

In some ways this seems to fill a need that spiritual beliefs once did, and still do, for many people. To a degree they smack of the religious principle of "faith without proof." They've provided us with assertions—teachings—by authority figures who know a lot more about these things than we do. They're doing "science," after all. And what they tell us is that a lot of it's pretty much beyond our control. We can't help the way we've evolved. If you want to blame anyone, blame our Paleolithic ancestors, whatever good that does.

The evidence for these assertions? Well, the scientists know a lot more about it. They have credentials based on years of research. We can still feel guilty if we want to, but what's the point?

Curtis White observes that "free market economists regularly reinforce the idea that economic markets are natural phenomena." As such, there's little that any would-be reformers can do to bring about more humane conditions. "If some people suffer for these facts, there is not much that can be done about it. In fact, there might even be some justice in their suffering (as in Ayn Rand)."[13] It might be a relief for some to hear that these things are totally out of our hands. The evolutionary psychologists (a.k.a., the "scientists") have made that pretty clear. A rather disturbing perspective, for sure.

Leo Panitch and Sam Gindin express a somewhat different view in their extensive analysis of global capitalism, however. "Political scientists, for their part, have usually understood that markets are not natural but need to be made, and that states are fundamental actors in this process."[14] *Capital*, by Thomas Piketty, which became a spectacular best seller in 2014, makes much the same point.[15]

It's really not a matter of social or economic systemic factors that we'd like to change but really can't—things like economic inequality, malnutrition, gender discrimination, high crime rates, or crooked political systems. We can change them. But not without concerted, collective effort, self-sacrifice, a lot of frustration, and a good chance of failure.

Changing behavior can be difficult, especially on a collective basis. But how many thousands of times has it happened nonetheless, in the course of the human experience? Generally, it happens through communication. That's how we've evolved—to be able to function that way, collectively, through communication. How ironic, at a time when instantaneous communication has become far easier than ever before in human history, giving rise to massive political movements that often surprise even professional observers, that we should be talking about "fixed behavior" patterns "selected for" in the Pleistocene.

The biological determinists are right in the assertion that many things about human life are out of our hands. We can't change our genes—at least, not at this point. But it's a major leap to assert that these unchangeable genes direct our behavior and that therefore, we can't change that either. History, anthropology, and the other social sciences and humanities—disciplines that some evolutionary psychologists find so unscientific and imprecise—emphatically tell us otherwise.

Notes

INTRODUCTION. WHERE ARE WE GOING WITH THIS?

1. Mal Ahern and Moira Weigel, "Survival of the Sexiest: How Evolutionary Psychology Went Viral," *The Nation*, Sept. 29, 2014, 17-22.

2. A salute to Marshall Sahlins, whose *Use and Abuse of Biology: An Anthropological Critique of Sociobiology* (Ann Arbor: University of Michigan Press, 1976) is a foundational work on this issue.

3. *Merriam-Webster Collegiate Dictionary*, 11th ed., s.v. "determinative."

4. Gunnar Myrdal, *An American Dilemma*, vol. 1, *The Negro Problem and Modern Democracy* (1944; repr. New York: HarperCollins, 1962), 83.

5. Noam Chomsky, *Aspects of the Theory of Syntax* (Cambridge, MA: MIT Press, 1965). We should note that many linguists have taken issue with this assertion. See, e.g., William A. Foley, "Do Humans Have Innate Mental Structures? Some Arguments from Linguistics," in Susan McKinnon and Sydel Silverman, eds., *Complexities: Beyond Nature and Nurture* (Chicago: University of Chicago Press, 2005), 43-63.

6. Noam Chomsky, *Language and Responsibility* (Hassocks: Harvester Press, 1979), 66.

7. Roy Edgley et al., "Noam Chomsky: An Interview," *Radical Philosophy* 53 (Autumn 1989): 31-40, 31.

8. See, e.g., Richard Dawkins, *The Selfish Gene* (Oxford: Oxford University Press, 1976).

9. See, e.g., Claude Lévi-Strauss, *The Elementary Structures of Kinship* (1949; repr. Boston: Beacon Press, 1969); *Structural Anthropology* (1958; repr. New York: Basic Books, 1963).

10. See Patrice Maniglier, "Claude Lévi-Strauss, 1908-2009: A Lévi-Straussian Century," *Radical Philosophy* 160 (Mar./Apr. 2010): 65-68.

11. Many, of course, would not consider Theodore Roosevelt especially liberal. There was that tendency to go to war, for example. But he also pushed domestic policies that favored the average citizen at the expense of big business.

12. Gregory Clark, "Your Ancestors, Your Fate," *New York Times*, Feb. 21, 2014.

13. Nicholas Wade, *A Troublesome Inheritance: Genes, Race, and Human History* (New York: Penguin, 2014).

14. H. Allen Orr, "Stretch Genes," *New York Review of Books*, June 5, 2014, 18-20.

15. *The Progressive*, July/Aug. 2014, 6.

CHAPTER 1. DON'T GET ME STARTED

1. Clyde C. Kluckhohn, *Anthropology and the Classics* (Providence, RI: Brown University Press, 1961), 34.

2. Herodotus, *The Histories* (New York: Oxford University Press, 1998), 590.

3. *Hippocrates: With an English Translation by W. H. S. Jones* (Cambridge, MA: Harvard University Press, 1995), 1:113.

4. Augustine ("Aurelius Augutinus") (350-430), *De Civite Dei* [The city of God], 16.9. Quoted in J. S. Slotkin, ed., *Readings in Early Anthropology*, Viking Fund Publications in Anthropological Research (New York: Wenner-Gren Foundation for Anthropological Research, 1965), 2-3.

5. Thomas Aquinas (1225-1274), *Summa Theologica*, 2.2.104. Quoted in Slotkin, *Readings in Early Anthropology*, 25.

6. Ibn Khaldun, *The Muqaddinmah: An Introduction to History* (New York: Pantheon, 1958), 174.

7. Ibn Khaldun, *The Muqaddinmah*, 282.

8. M. C. Seymour, ed., *Mandeville's Travels* (Oxford: Clarendon, 1967), xix.

9. Quoted in John Howland Rowe, "The Renaissance Foundations of Anthropology," *American Anthropologist* 67 no. 1 (1965): 1-20, reprinted in Regna Darnell, *Readings in the History of Anthropology* (New York: Harper & Row, 1974), 61-77, 75.

10. Michel de Montaigne, *Essays*, reprinted in Darnell, *Readings in the History of Anthropology*, 128-139, 134.

11. Montaigne, *Essais* III,142. Quoted in Slotkin, *Readings in Early Anthropology*, 56.

12. Leonardo da Vinci, *Notebooks*. Quoted in Slotkin, *Readings in Early Anthropology*, 39.

13. Baron de Montesquieu, *The Spirit of the Laws*, reprinted in Darnell, *Readings in the History of Anthropology*, 157-168, 158, 159.

14. Montesquieu, *The Spirit of the Laws*, reprinted in Darnell, *Readings in the History of Anthropology*, 160.

15. Andreas Vesalius, *De humani corporis fabrica*, 23. Quoted in Slotkin, *Readings in Early Anthropology*, 40.

16. Jean Bodin, *Des six livres*, 698-699. Quoted in Slotkin, *Readings*, 71.

17. William Robertson, *The History of America* (1777), reprinted in Darnell, *Readings in the History of Anthropology*, 140-151, 142.

18. See, e.g., Fritz Graebner's culture-circle theories, "Kulturkreise und Kulturschichten in Ozeanien," *Zeitschrift für Ethnologie*, 37 (1903), 28-53, discussed in Marvin Harris, *The Rise of Anthropological Theory* (New York: Thomas Y. Crowell, 1968), 383-384.

19. Carolus Linnaeus, *Systema Naturae* (1735).

20. See discussion in Richard J. Perry, *"Race" and Racism: The Development of Modern Racism in America* (New York: Palgrave Macmillan, 2007), 127-128.

21. See Paul Lawrence Farber, *Finding Order in Nature: The Naturalist Tradition from Linnaeus to E. O. Wilson* (Baltimore: Johns Hopkins University Press, 2000), 21, for a discussion of this.

22. Farber, *Finding Order in Nature*, 104.

23. See Paul Lawrence Farber, *Mixing Races: From Scientific Racism to Modern Evolutionary Ideas* (Baltimore: Johns Hopkins University Press, 2011), 28, for a discussion of this.

24. Johann Friedrich Blumenbach, *On the Natural Varieties of Mankind: De Generis Human Varietate Nativa* (New York: Bergman, 1969). Discussed in Bruce Dain, *A Hideous Monster of the Mind: American Race Theory in the Early Republic* (Cambridge, MA: Harvard University Press, 2002), 59.

25. See Farber, *Finding Order*, 21-33.

26. See Henry M. Stanley, *In Darkest Africa and the Quest, Rescue, and Retreat of Emin Governor of Equatoria* (New York: Charles Scribner's Sons, 1891), 228-229, 431.

27. Josiah Clark Nott, "Hybridity of Animals, Viewed in Connection with the Natural History of Mankind," in Josiah Clark Nott and George R. Gliddon, eds., *Types of Mankind: Or Ethnological Researches, Based upon the Ancient Monuments, Paintings, Sculptures, and Crania of Races* (Philadelphia: J. B. Lippincott, 1865), 372-410, 373.

28. See, e.g., Richard Hofstadter, *Social Darwinism in American Thought* (1944; repr. New York: George Braziller, 1955), 172.

29. The Linguistic Society of America offers the estimate of 6,909 as of 2009. Linguistic Society of America, "How Many Languages Are There in the World?," http://www.linguisticsociety.org/content/how-many-languages-are-there-world.

CHAPTER 2. EUGENICS

1. J. R. Miller, *Skyscrapers Hide the Heavens: A History of Indian-White Relations in Canada* (1989; repr. Toronto: University of Toronto Press, 1991), 52.

2. See Richard J. Perry, *From Time Immemorial: Indigenous Peoples and State Systems* (Austin: University of Texas Press, 1996), 95.

3. See William Christie MacLeod, "Celt and Indian: Britain's Old World Frontier in Relation to the New," in Paul Bohannan and Fred Ploeg, eds., *Beyond the Frontier: Social Process and Culture Change* (Garden City, NY: Natural History Press, 1967), 25-41; Audrey Smedley, *Racism in North America: Origins and Evolution of a World View* (Boulder, CO: Westview, 1999), 53; Christine Bolt, *American Indian Policy and American Reform: Case Studies of the Campaign to Assimilate the American Indian* (London: Allen and Unwin, 1987).

4. Edwin Black, *War against the Weak: Eugenics and America's Campaign to Create a Master Race* (New York: Four Walls Eight Windows, 2003), 186.

5. See Michael Burawoy, "The Function and Reproduction of Migrant Labor: Comparative Material from Southern Africa and the United States," *American Journal of Sociology* 81 (1976): 1050-1087.

6. See David Ray Papke, *The Pullman Case: The Clash of Labor and Capital in Industrial America*. Lawrence: University of Kansas Press, 1999.

7. Thomas Jefferson, *Notes on the State of Virginia* (1781-1785; repr. Chapel Hill: University of North Carolina Press, 1955), 18.

8. Lothrop Stoddard, *The Rising Tide of Color against White World Supremacy* (New York: Charles Scribner's Sons, 1926), 259-260, 306. Quoted in Black, *War against the Weak*, 133.

9. Quoted in Black, *War against the* Weak, 133.

10. Quoted in Black, *War against the Weak*, 65.

11. Quoted in Black, *War against the Weak*, 138.

12. Harry Laughlin, "Calculations on the Working Out of a Proposed Program of Sterilization," paper read at the First National Conference on Race Betterment, Battle Creek, Michigan, 1912. Quoted in Black, *War against the Weak*, 88.

13. Herbert Spencer, *Principles of Sociology* (New York: Appleton-Century-Crofts, 1883).

14. Quoted in Black, *War against the Weak*, 29.

15. Margaret Sanger, *The Pivot of Civilization* (New York: Brentano's, 1922). Quoted in Black, *War against the Weak*, 129.

16. John C. Duvall, "The Purpose of Eugenics," *Birth Control Review*, Dec. 1924, 345.

17. Black, *War against the Weak*, 58.

18. Cited in Daniel J. Kevles, *In the Name of Eugenics: Genetics and the Uses of Human Heredity* (New York: Alfred A. Knopf, 1985), 49.

19. Raymond Pearl, "The Biology of Superiority," *American Mercury* 12 (Nov. 1927): 260.

20. Black, *War against the Weak*, 133.

21. Black, *War against the Weak*, 73.

22. See, e.g., Madison Grant, *The Passing of the Great Race* (New York: Charles Scribner's Sons, 1916); Stoddard, *Rising Tide of Color*.

23. Black, *War against the Weak*, 259.

24. Black, *War against the Weak*, 40, 45.

25. Black, *War against the Weak*, 277.

26. See the Southern Poverty Law Center's *Intelligence Report* for annual reports on these groups.

27. Sanger, *Pivot of Civilization*, 101-102; Sanger, *An Autobiography* (New York: W. W Norton, 1938), 376-377. Discussed in Black, *War against the Weak*, 128-133.

28. Rob DeSalle and Michael Yudell, *Welcome to the Genome: A User's Guide to the Past, Present, and Future* (New York: Wiley, 2005), 17.

29. Hilary Rose and Steven Rose, *Genes, Cells, and Brains: The Promethean Promises of the New Biology* (New York: Verso, 2012), 127.

30. Jim Adams, "Sterilization Program Targeted Abenaki," *Indian Country Today* (Lakota Times), Feb. 4, 2004.

31. Black, *War against the Weak*, 1ff.

32. Susan Donaldson James and Courtney Hutchinson, "N.C. to Compensate Victims of Sterilization in 19th Century Eugenics Program," *ABC News*, Jan. 10, 2012.

33. J. Haldane, "Lysenko and Genetics," *Science and Society* 4 (1940): 4.

34. Alfred Binet and Theodore Simon, *The Development of Intelligence in Children (The Binet-Simon Scale)*, Elizabeth S. Kite, trans. (1916; repr. Nashville, TN: Williams Printing, 1920).

35. Henry H. Goddard, "The Binet Tests in Relation to Immigration," *Journal of Psycho-Asthenics* 18, no. 12 (1913), 105-109; "Mental Tests and the Immigrants," *Journal of Delinquency* 2, no. 5 (1917), 243-277; Lewis M. Terman, *The Measurement of Intelligence: An Explanation and Complete Guide for the Use of the Stanford Revision and Extension of the Binet-Simon Intelligence Scale* (Boston: Houghton Mifflin, 1916).

36. See Stephen Jay Gould, *The Mismeasure of Man* (New York: Norton, 1981). See also David H. Price, *Threatening Anthropology: McCarthyism and the FBI, Surveillance of Activist Anthropologists* (Durham, NC: Duke University Press, 2004), 128.

37. Quoted in Kevles, *In the Name of Eugenics*, 139.

38. Quoted in Kevles, *In the Name of Eugenics*, 129.

39. Walter Lippmann, "The Great Confusion: A Reply to Mr. Terman," *New Republic* (Jan. 3, 1923): 145-146.

40. See, however, Thomas Kuhn, *The Structure of Scientific Revolutions* (Chicago: University of Chicago Press, 1970), on the persistence of scientific paradigms.

41. Franz Boas, *Abstracts on the Report of Changes in Bodily Form of Descendants of Immigrants* (Washington, DC: Government Printing Office, 1912).

42. The geneticist Edward M. East of Harvard in the immediate aftermath of Boas's findings wrote that "today the Jews retaliate by proclaiming that the Nordic race is a myth." Dr. Ellsworth Huntington dismissed Boas as representing "a relatively small group of scientific men, especially those who belong to races that are not dominant." Quoted in Ashley Montagu, *Frontiers of Anthropology* (New York: Perogee, 1974), 443.

43. See Garland Allen, "Is a New Eugenics Afoot?," *Science* 294, no. 5540 (Oct. 5, 2001): 60.

44. Corey Sparks and Richard J. Jantz, "Changing Times, Changing Faces: Franz Boas's Immigrant Study in Modern Perspective," *American Anthropologist* 105, no. 2 (2003): 333–337.

45. Nicholas Wade, "A New Look at Old Data May Discredit a Theory on Race," *New York Times*, Oct. 8, 2002.

46. Charles C. Gravlee, H. Russell Bernard, and William A. Leonard, "Boas's Changes in Bodily Form: The Immigrant Study, Cranial Plasticity, and Boas's Physical Anthropology," *American Anthropologist* 105, no. 2 (2003): 326–332.

47. Lee D. Baker, "The Cult of Franz Boas and His "Conspiracy" to Destroy the White Race," *Proceedings of the American Philosophical Society* 154, no. 1 (2010): 8–18.

48. See Nicholas Lemann, *The Big Test: The Secret History of the American Meritocracy* (New York: Farrar, Straus and Giroux, 1999), for a discussion of the growth of the testing industry in the United States.

49. See James R. Flynn, "Massive Gains in IQ in 14 Nations: What IQ Tests Really Measure," *Psychological Bulletin* 101 (1987): 171–191.

50. Claude S. Fischer, "Are Humans Getting Smarter?," *Boston Review*, June 3, 2013.

3. KILLER APES, NAKED APES, AND JUST PLAIN NASTY PEOPLE

1. Many writers now prefer the term *hominin* to refer specifically to the human line rather than the older term *hominid*, which once had the same meaning but now has come to include apes and their ancestors.

2. Robert Ardrey, *African Genesis: A Personal Investigation into the Animal Origins and Nature of Man* (New York: Dell, 1961).

3. Ardrey, *African Genesis*, 299.

4. See, e.g., C. K. Brain, "New Finds at the Swartkrans Site," *Nature* 225 (1970): 1112–1119.

5. Clifford J. Jolly, "The Seed-Eaters: A New Model of Hominid Differentiation Based on a Baboon Analogy," *Man*, n.s., 5, no. 1 (1970): 5–26; Richard F. Kay, "Dental Evidence for the Diet of Australopithecines," *Annual Review of Anthropology* 14 (1985): 315–341.

6. See, e.g., Frank Speck, *Naskapi: The Savage Hunters of the Labrador Peninsula* (Norman: University of Oklahoma Press, 1977); Adrian Tanner, *Bringing Home Animals: Religious Ideology and Mode of Production of the Mistassini Cree Hunters* (New York: St. Martin's Press, 1979); Richard J. Perry, *Western Apache Heritage: People of the Mountain Corridor* (Austin: University of Texas Press, 1991), 32.

7. Robert Ardrey, *The Territorial Imperative: A Personal Inquiry into the Animal Origins of Property and Nations* (New York: Dell, 1966).

8. Perry, *Western Apache Heritage*, 26-27.

9. Desmond Morris, *The Naked Ape: A Zoologist's Study of the Human Animal* (New York: McGraw-Hill, 1967).

10. Konrad Lorenz, *On Aggression* (1966; repr. New York: Bantam Books, 1970).

11. Lorenz, *On Aggression*. Excerpt reprinted as "The Functional Limits of Morality," in Arthur L. Caplan, ed., *The Sociobiology Debate: Readings on Ethical and Scientific Issues* (New York: Harper & Row, 1978), 67-75, 67.

12. Lorenz, *On Aggression*, 70.

13. Lorenz, *On Aggression*, 70.

14. Lorenz, *On Aggression*, 71.

15. Lorenz, *On Aggression*, 70.

16. Lorenz, *On Aggression*, 71.

17. Lorenz, *On Aggression*, 71.

18. Lorenz, *On Aggression*, 72.

19. Lorenz, *On Aggression*, 72.

20. Konrad Lorenz, "On the Foundations of the Concept of Instinct," *Natural Sciences* 25, no. 29 (1937): 289-300.

21. See, e.g., Alexander Alland, *Race in Mind: Race, IQ, and Other Racisms* (New York: Palgrave, 2002); C. Loring Brace, *"Race" Is a Four-Letter Word: The Genesis of the Concept* (New York: Oxford University Press, 2005); Richard C. Lewontin, Steven P. R. Rose, and Leon J. Kamin, *Not in Our Genes: Biology, Ideology, and Human Nature* (New York: Pantheon Books, 1984); Jonathan Marks, *Human Biodiversity: Genes, Race, and History* (New York: Aldine de Gruyter, 1995).

22. Richard C. Lewontin, "The Apportionment of Human Diversity," *Evolutionary Biology* 6 (1972): 381-398; Lewontin, Rose, and Kamin, *Not in Our Genes*.

23. L. Luca Cavalli-Sforza and Walter F. Bodmer, *The Genetics of Human Populations* (San Francisco: W. H. Freeman, 1971).

24. Marks, Human Biodiversity, 130.

25. See, e.g., Carleton S. Coon, *The Origin of Races* (New York: Knopf, 1962).

26. See Alan R. Templeton, "Human Races: A Genetic and Evolutionary Perspective," in Robert W. Sussman, ed., *The Biological Basis of Human Behavior: A Critical Review* (Upper Saddle River, NJ: Prentice Hall, 1999), 180-192.

CHAPTER 4. MIND GAMES

1. Stephen Jay Gould, *The Mismeasure of Man* (New York: Norton, 1981), 24-25.

2. See, e.g., Craig T. Ramey, "High-Risk Children and IQ: Altering Intergenerational Patterns," *Intelligence* 16 (1992): 239-256.

3. Arthur J. Jensen, "How Much Can We Boost IQ and Scholastic Achievement?," *Harvard Educational Review* 39(1969): 1-123.

4. *People,* May 1979. See also omg-facts.com.

5. See L. J. Kamin, *The Science and Politics of I.Q.* (New York: Wiley, 1974); L. S. Hearnshaw, *Cyril Burt, Psychologist* (Ithaca, NY: Cornell University Press, 1979).

6. U.S. Office of Health and Human Services, *Third Grade Follow-Up to the Head Start Impact Study: Final Report* (Rockville, MD: Office of Planning, Review, and Evaluation, Dec. 21, 2012).

7. See Rick Weiss, "Beyond Nature vs. Nurture: Positive Environmental Factors Can Boost Poorer Students' IQ Scores, Researchers Show," *Washington Post National Weekly Edition,* Sept. 8–14, 2005.

8. Richard Herrnstein, "I.Q.," *Atlantic Monthly* 228 (1971): 43–64.

9. Richard Herrnstein and Charles Murray, *The Bell Curve: Intelligence and Class Structure in American Society* (New York: Free Press, 1994).

10. Gina Kolata, "Study Raises the Estimates of Inherited Intelligence," *New York Times,* Oct. 12, 1990.

11. Nicholas Wade, "A New Look at Old Data May Discredit a Theory on Race," *New York Times,* Oct. 8, 2002. Wade published a good deal promoting the idea of genetic influence on behavior, including his book *The Faith Instinct* (New York: Penguin, 2009) and *A Troublesome Inheritance: Genes, Race and Human History* (New York: Penguin, 2014).

12. See, e.g., Lee D. Baker, "The Cult of Franz Boas and His 'Conspiracy' to Destroy the White Race," *Proceedings of the American Philosophical Society* 154, no. 1 (2010): 8–18.

13. For a classic discussion of this, see Mary Douglas, *Purity and Danger: An Analysis of Concepts of Pollution and Taboo* (London: Routledge and Kegan Paul, 1966).

14. Thomas J. Bouchard Jr., David T. Lykken, Matthew McGue, Nancy L. Segal, and Auke Tellegen, "Sources of Human Psychological Differences: The Minnesota Study of Twins Reared Apart," *Science* 250 (1990): 223–228.

15. Frank J. Sulloway, "Parallel Lives," *New York Review of Books,* Nov. 30, 2006. See also a response: Jack Kaplan, "How to Inherit IQ: An Exchange," *New York Review of Books,* Mar. 15, 2007.

16. See, e.g., David S. Moore, *The Dependent Gene: The Fallacy of "Nature vs. Nurture"* (New York: Henry Holt/Times Books, 2003).

17. See Ramey, "High-Risk Children and I.Q."

18. Mario F. Fraga et al., "Epigenetic Differences Arise during the Lifetime of Monozygotic Twins," *Proceedings of the National Academy of Sciences* 102 (2005): 10604–10609.

19. Rick Weiss, "Twin Data Highlights Genetic Changes," *Washington Post,* July 5, 2005, 2. Quoted in Susan McKinnon, *Neo-liberal Genetics: The Myths and Moral Tales of Evolutionary Psychology* (Chicago: Prickly Paradigm Press, 2005), 26.

20. J. Philippe Rushton, *Race, Evolution, and Behavior: A Life History Perspective* (New Brunswick, NJ: Transaction, 1995).

21. Hans J. Eysenck, *The IQ Argument: Race, Intelligence, and Education* (New York: Library Press, 1971).

22. Hans J. Eysenck, "Science, Racism, and Sexism," *Journal of Political, Social, and Economic Studies* 16 (1981): 215-250, 217.

23. J. Philippe Rushton and Arthur J. Jensen, "Thirty Years of Research of Race Differences in Cognitive Ability," *Psychology*, Public Policy and Law 11, no. 2 (2005): 235-294.

24. "Pioneer Fund," Southern Poverty Law Center, www.splcenter.org/get -informed/intelligence-files/groups/pioneer-fund.

25. The Pioneer Fund, www.thepioneerfund.org.

CHAPTER 5. SOCIOBIOLOGY

1. S. H. Waddington, "Mindless Societies," *New York Review of Books*, Aug. 7, 1975, 252-258, 252.

2. E. O. Wilson, *Sociobiology: The New Synthesis* (Cambridge, MA: Harvard University Press, 1975).

3. Reprinted in Arthur L. Caplan, *The Sociobiology Debate: Readings on Ethical and Scientific Issues* (New York: Harper and Row, 1978), 84.

4. Richard Dawkins, *The Selfish Gene* (Oxford: Oxford University Press, 1989).

5. Waddington, "Mindless Societies."

6. See Susan McKinnon, *Neo-liberal Genetics: The Myths and Moral Tales of Evolutionary Psychology* (Chicago: Prickly Paradigm Press, 2005), 77-79, for a concise discussion of this.

7. Wilson, *Sociobiology*, 549.

8. Wilson, *Sociobiology*, 559.

9. Wilson, *Sociobiology*, 559.

10. For an excellent discussion and critique of sociobiologists' use of anthropomorphism, see Philip Kitcher, *Vaulting Ambition: Sociobiology and the Quest for Human Nature* (Cambridge, MA: MIT Press, 1985), 183-201.

11. Randy Thornhill and John Alcock, *The Evolution of Insect Mating Systems* (Cambridge: Harvard University Press, 1983).

12. The idea of rape as a biological strategy arose again many years later in the context of evolutionary psychology, which we'll discuss in chapter 6. Randy Thornhill and Craig Palmer, *A Natural History of Rape: Biological Bases of Sexual Coercion* (Cambridge, MA: MIT Press, 2000). In this case, the exemplars were mallard ducks and scorpion flies. See Robert C. Richardson, *Evolutionary Psychology as Maladapted Psychology* (Cambridge, MA: MIT Press, 2007), 36-37, for a brief critique, and Kitcher, *Vaulting Ambition*, 185-189, for a more extensive critical discussion.

13. Wilson, *Sociobiology*, 4.

14. Nor, unfortunately, has it stopped others from pursuing the same rather treacherous path. In 2012, the prolific ornithologist Jared Diamond published a book arguing that we would benefit from emulating the ways of our ancestors, as carried on among nonindustrial peoples. The title of the book, which is based on Diamond's many years of experience living in Papua New Guinea, is *The World until Yesterday: What We Can Learn from Tribal Societies* (New York: Viking, 2012).

15. For discussion of this, see C. Loring Brace, "Australian Tooth Size and the Death of a Stereotype," *Current Anthropology* 24, no. 2 (1980): 141-164.

16. See, e.g., Richard B. Lee, *The !Kung San: Men, Women, and Work in a Foraging Society* (New York: Cambridge University Press, 1979).

17. Wilson, *Sociobiology*, 549.

18. See Marjorie Shostak, *Nisa: The Life and Worlds of a !Kung Woman* (New York: Vintage Books, 1981).

19. See Robert Trivers, "Parental Investment and Sexual Selection," in B. Campbell, ed., *Sexual Selection and the Descent of Man* (Chicago: Aldine, 1972).

20. See McKinnon, *Neo-liberal Genetics*, 77-79, for a good discussion of this.

21. Dan Slater, "Darwin Was Wrong about Dating," *New York Times*, Jan. 12, 2013.

22. Kate Taylor, "She Can Play That Game, Too," Sunday Styles, *New York Times*, July 14, 2013.

23. Po Bronson and Ashley Merryman, "Why Worry?," *New York Times Magazine*, Feb. 10, 2013, 20-25.

24. Bronson and Merryman, "Why Worry?," 24.

CHAPTER 6. AND YET ANOTHER NEW SCIENCE
OF THE SAME OLD THING

1. Steven Pinker, *The Blank Slate: The Modern Denial of Human Nature* (New York: Viking, 2002), 241.

2. Pinker, *Blank Slate*, 241.

3. For a psychologist's view on this, see Jefferson M. Fish, "Why Psychologists Should Learn Some Anthropology," *American Psychologist* 50, no. 1 (1995): 44-45; Fish, "What Anthropology Can Do for Psychology: Facing Physics Envy, Ethnocentrism, and a Belief in 'Race,'" *American Anthropologist* 102, no. 3 (2000): 552-563; Fish, "How Psychologists Think about 'Race,'" *General Anthropology* 4, no. 1 (1997): 1-4.

4. See, e.g., Ruth Benedict, *Patterns of Culture* (New York: New American Library, 1934); Abram Kardiner and Ralph Linton, *The Individual and His Society: The Psychodynamic of Primitive Social Organization* (New York: Columbia University Press, 1939); Robert A. LeVine, ed., *Culture and Personality: Contemporary Readings* (New York: Aldine, 1974).

5. Leda Cosmides and John Tooby, "Cognitive Adaptations for Social Exchange," in Jerome H. Barkow, H. L. Cosmides, and John Tooby, eds., *The Adapted Mind: Evolutionary Psychology and the Generation of Culture* (Oxford: Oxford University Press, 1992), 163-228, 163.

6. Paul Seabright, *The War of the Sexes: How Conflict and Cooperation Have Shaped Men and Women from Prehistory to the Present* (Princeton, NJ: Princeton University Press, 2012).

7. However, one researcher in 2013 concluded that competition and hostility among women (a.k.a. the "mean girls" phenomenon) is also genetically based behavior, selected for in the Pleistocene as a means of competing for the most desirable mates. Gia Ghose, "Mean Girls: Women Evolved to Be Catty?," *Livescience*, Oct. 27, 2013.

8. Robert L. Trivers, "The Evolution of Reciprocal Altruism," *Quarterly Review of Biology* 46 (1971): 35-57; Trivers, "Parental Involvement in Sexual Selection," in Bernard G. Campbell, ed., *Sexual Selection and the Descent of Man* (Chicago: Aldine, 1972).

9. Kate Taylor, "She Can Play That Game, Too," *New York Times*, Jan. 12, 2013, 6.

10. Andreas Olasson, Jeffrey P. Ebert, Mahzarin R. Banaji, and Elizabeth A. Phelps, "The Role of Social Groups in the Persistence of Learned Fear," *Science* 309 (2005): 785-787.

11. Olasson et al., "Role of Social Groups," 787.

12. Chris Mooney, *The Republican Brain: The Science of Why They Deny Science—and Reality* (New York: Wiley, 2012).

13. Pinker, *Blank Slate*, 144.

14. Pinker, *Blank Slate*, 106-107.

15. Cf. Pinker, *Blank Slate*, 110-111.

16. Pinker, *Blank Slate*, ix.

17. Pinker, *Blank Slate*, x.

18. Pinker, *Blank Slate*, x.

19. Pinker, *Blank Slate*, 157-158.

20. Émile Durkheim, *The Rules of the Sociological Method* (1895; repr. New York: Free Press, 1938).

21. David M. Buss, *Evolutionary Psychology: The New Science of the Mind* (New York: Allyn and Bacon, 2012), 12.

22. Jaime C. Confer, Judith A. Easton, Diana S. Fleischman, Carl D. Goetz, David M. G. Lewis, Carin Perilloux, and David M. Buss, "Evolutionary Psychology: Controversies, Questions, Prospects and limitations," *American Psychologist* 65, no. (2010): 110-126, 114.

23. Confer et al., "Evolutionary Psychology," 373.

24. Confer et al., "Evolutionary Psychology," 68.

25. Confer et al., "Evolutionary Psychology," 19.

26. From Susan McKinnon, *Neo-liberal Genetics: The Myths and Moral Tales of Evolutionary Psychology* (Chicago: Prickly Paradigm Press, 2005), 30-31.

27. Steven Pinker, *How the Mind Works* (New York: W. W. Norton, 1997).

28. Confer et al., "Evolutionary Psychology," 120.

29. *The American Heritage Dictionary of the English Language*, William Morris, ed. (Boston: American Heritage and Houghton Mifflin, 1973).

30. See, e.g., Daniel Bergner, *What Do Women Want? Adventures in the Science of Female Desire* (New York: Ecco/HarperCollins, 2013).

CHAPTER 7. THAT'S JUST ABOUT ENOUGH OF THAT

1. Steven Pinker, *The Blank Slate: The Modern Denial of Human Nature* (New York: Viking, 2002), x.

2. Susan McKinnon and Sydel Silverman, eds., *Complexities: Beyond Nature and Nurture* (Chicago: University of Chicago Press, 2005), 17.

3. Pinker, *Blank Slate*, 421.

4. Pinker, *Blank Slate*, 38.

5. Richard A. Shweder, "You're Not Sick, You're Just in Love: Emotion as an Interpretative System," in Paul Ekman and Richard J. Davidson, eds., *The Nature of Emotion: Fundamental Questions* (New York: Oxford University Press, 1994), 32-44.

6. Shweder, "You're Not Sick," 32-44.

7. With apologies to John Dryden (1669-1670).

8. Pinker, *Blank Slate*, 230.

9. Pinker, *Blank Slate*, 23.

10. Pinker, *Blank Slate*, 272.

11. See Margaret Caffrey, *Ruth Benedict: Stranger in This Land* (Austin: University of Texas Press, 1989).

12. Melville J. Herskovits, *The Myth of the Negro Past* (New York: Harper and Brothers, 1941).

13. Zora Neale Hurston, *Dust Tracks on the Road: An Autobiography*. (Urbana: University of Illinois Press, 1970).

14. Pinker, *Blank Slate*, 9.

15. Alfred L. Kroeber, "The Superorganic," *American Anthropologist* 19 (1917): 163-213. Herbert Spencer had developed a similar concept of society, based on his "organic analogy." Herbert Spencer, *Principles of Sociology* (New York: Appleton-Century-Crofts, 1888). As discussed below, however, Spencer's perspective on the nature of society was quite different from Kroeber's.

16. Pinker, *Blank Slate*, ch. 5.

17. Pinker, *Blank Slate*, 255.

18. Pinker, *Blank Slate*, 319.

19. See, e.g., Hilary Rose and Steven Rose, *Genes, Cells, and Brains: The Promethean Promises of the New Biology* (New York: Verso, 2012), 213.

20. Kathleen R. Gibson, "Epigenesis, Brain Plasticity, and Behavioral Versatility: Alternatives to Standard Evolutionary Psychology Models," in McKinnon and Silverman, *Complexities*, 23-42, 28.

21. Gibson, "Epigenesis," 28.

22. See Marlene Zuk, *Paleofantasy: What Evolution Really Tells Us about Sex, Diet, and How We Live* (New York: W. W. Norton, 2013).

23. See, e.g., Sam Stein, "Sequestration Pushes Head Start Families to the Precipice," *Huffington Post*, July 9, 2013.

24. See, e.g., US Department of Health and Human Services, *Third Grade Follow-Up to the Head Start Impact Study: Final Report* (Washington, DC: Office of Planning, Review, and Evaluation, Dec. 21, 2012).

25. Thanks to Jaime C. Confer, Judith A. Easton, Diana S. Fleischman, Carl D. Goetz, David M. G. Lewis, Carin Perilloux, and David M. Buss, "Evolutionary Psychology: Controversies, Questions, Prospects and limitations," *American Psychologist* 65, no. (2010): 113, for this summary.

26. Michael E. Price, Leda Cosmides, and John Tooby, "Punitive Sentiment as an Anti-Free-Rider Psychological Device," *Evolution and Human Behavior* 23 (2002): 203-231.

27. Leda Cosmides and John Tooby, "Neurocognitive Adaptations Designed for Social Exchange," in David M. Buss, ed., *Handbook of Evolutionary Psychology* (New York: Wiley, 2005), 584-627.

28. L. Silverman and J. Choi, "Locating Places," in Buss, *Handbook of Evolutionary Psychology*, 177-199.

29. Heidi Greiling and David M. Buss, "Women's Sexual Strategies: The Hidden Dimension of Short-Term Extra-Pair Mating," *Personality and Individual Differences* 28 (2000): 929-963; D. P. Schmidt and International Sexuality Description Project, "Universal Sex Differences in the Desire for Sexual Variety: Tests from 52 Nations, 6 continents, and 13 Islands," *Journal of Personality and Social Psychology* 85 (2004): 85-104.

30. M. G. Hazelton, D. M. Buss, V. Oubaid, and A. Angleitner, "Sex, Lies, and Strategic Interference: The Psychology of Deception between the Sexes," *Personality and Social Psychology Bulletin* 31 (2005): 3-23.

31. Confer et al., "Evolutionary Psychology," 113.

32. Rose and Rose, *Genes, Brains, and Cells*, 82.

33. See Marlene Zuk, *Paleofantasy*, esp. 104-106.

34. See Robert L. Trivers, "Parental Investment and Sexual Selection," in Bernard G. Campbell, ed., *Sexual Selection and the Descent of Man* (Chicago: Aldine, 1972).

35. Dan Slater, "Darwin Was Wrong about Dating," *New York Times*, Jan. 12, 2013.

36. Kate Taylor, "She Can Play That Game, Too," *New York Times*, July 14, 2013.

37. Confer et al., "Evolutionary Psychology," 113, 122.

38. See Susan McKinnon, *Neo-liberal Genetics: The Myths and Moral Tales of Evolutionary Psychology* (Chicago: Prickly Paradigm Press, 2005), 162.

39. Buss, *Handbook of Evolutionary Psychology*, 18.

40. Antonia Abbey, "Sex Differences in Attribution of Friendly Behavior: Do Males Misperceive Females?," *Journal of Personality and Social Psychology* 43 (1982): 830–838, 832.

41. See McKinnon, *Neo-liberal Genetics*, for a discussion of this.

42. Richard B. Lee, *The !Kung San: Men, Women, and Work in a Foraging Society* (Cambridge: Cambridge University Press, 1979).

43. Thomas Hobbes, *The Leviathan*, Richard Tuck, ed. (New York: Cambridge University Press, 1996).

44. Pinker, *Blank Slate*, 6–8.

45. For a lengthy discussion of this, see Robert C. Richardson, *Evolutionary Psychology as Maladapted Psychology* (Cambridge, MA: MIT Press, 2007), 41–88.

CHAPTER 8. IT'S NOT THAT SIMPLE

1. Hilary Rose and Steven Rose, *Genes, Cells, and Brains: The Promethean Promises of the New Biology* (New York: Verso, 2012), 1–2.

2. The Southern Poverty Law Center notes that from 2008 to 2013, so-called patriot and militia groups have increased in the United States from 149 to 1,360. *Intelligence Report* 149 (Spring 2013): 41.

3. William Faulkner, *Requiem for a Nun* (New York: Random House, 1951).

4. Rose and Rose, *Genes, Cells, and Brains*, 154.

5. Rose and Rose, *Genes, Cells, and Brains*, 213.

6. Herbert Jennings, *Prometheus: Or Biology and the Advancement of Man* (New York: E. P. Dutton, 1925), 17–19.

7. Daniel J. Kevles, *In the Name of Eugenics: Genetics and the Uses of Human Heredity* (New York: Alfred A. Knopf, 1985), 145.

8. See Ajit Varki and David L. Nelson, "Genomic Comparison of Humans and Chimpanzees," *Annual Review of Anthropology* 36 (2007): 191–209.

9. Eric Lander, quoted in Rose and Rose, *Genes, Cells, and Brains*, 278.

10. Steven Pinker, *The Blank Slate: The Modern Denial of Human Nature* (New York: Viking, 2002), 6–8.

11. Susan McKinnon and Sydell Silverman 2010, 9.

12. Peter Kropotkin, *Mutual Aid: A Factor of Evolution* (1901; repr. London: Heinemann, 1919).

13. Bronisław Malinowski, *A Scientific Theory of Culture and Other Essays* (Chapel Hill: University of North Carolina Press, 1944).

14. Richard J. Perry, "Radcliffe-Brown and Kropotkin: The Heritage of Anarchism in British Social Anthropology," *Kroeber Anthropological Society Papers* 51 and 52 (1978): 61–65.

15. William Graham Sumner of Yale, a major proponent of social Darwinism, wrote early in the twentieth century that in his view, socialism was "any device whose aim is to save individuals from any of the difficulties or hardships of the struggle for existence and the competition of life by the intervention of the state." Alfred G. Keller and Maurice R. Davie, eds., *Essays of William Graham Sumner* (New Haven, CT: Yale University Press, 1934), 435. Kropotkin probably would have agreed with him about state intervention.

16. Richard Hofstadter, *Social Darwinism in American Thought* (1944; repr. New York: George Braziller, 1955), 50.

17. Ayn Rand, *Atlas Shrugged* (New York: Dutton, 1992).

18. Ayn Rand and Nathaniel Branden, *The Virtue of Selfishness* (New York: Signet, 1964).

19. Herbert Spencer, *Social Statics* (New York: D. Appleton, 1864), 414–415. Quoted in Hofstadter, *Social Darwinism*, 41.

20. Hofstadter, *Social Darwinism*, 41.

21. Jason Richwine, "IQ and Immigration Policy" (doctoral dissertation, Harvard University, 2009) (UMI 3385409).

CHAPTER 9. WHAT'S THE AGENDA?

1. Edwin Black, *War against the Weak: Eugenics and America's Campaign to Create a Master Race* (New York: Four Walls Eight Windows, 2003), 57.

2. Garland Allen, "Is a New Eugenics Afoot?," *Science* 294, no. 5540 (Oct. 5, 2001): 3.

3. Matt Welch, "Arizona GOP VP Resigns after Advocating Sterilization and Drug Testing for Medicaid Recipients," *Hit & Run* (blog), Reason.com, Sept. 14, 2014, http://reason.com/blog/2014/09/15/arizona-gop-vp-resigns-after-advocating. See also Zachary Roth, "Top Arizona GOPer Russell Pearce Resigns after Sterilization Comments," MSNBC.com, Sept. 15, 2014, http://www.msnbc.com/msnbc/top-arizona-goper-resigns-after-sterilization-comments, Sept. 15, 2014.

4. Author's personal experience.

5. Michael A. Fletcher, quoted in Susan Perry, "The Income Gap Plays Out in U.S. Life Expectancies," *MinnPost*, Mar. 19, 2013, http://www.minnpost.com/second-opinion/2013/03/income-gap-plays-out-us-life-expectancy.

6. Sarah Kliff, "The U.S. Ranks 26th for Life Expectancy, Right behind Slovenia," *Wonkblog, Washington Post*, Nov. 21, 2013, http://www.washingtonpost.com/blogs/wonkblog/wp/2013/11/21/the-u-s-ranks-26th-for-life-expectancy-right-behind-slovenia/

7. "Maternal Mortality a Big Issue for Women's Health in U.S.," Face the Facts USA, July 2013, http://www.facethefactsusa.org/facts/more-us-mothers-dying -despite-expensive-care.

8. Save the Children Foundation, quoted in Tara Culp-Ressler, "Report: U.S. Has the Highest First-Day Infant Death Rate in the Industrialized World," *ThinkProgress*, May 7, 2013, http://thinkprogress.org/health/2013/05/07/1973341/us-infant-mortality -rate/

9. Herbert Spencer, *Social Statics; or, the Conditions Essential to Human Happiness Specified, and the First of Them Developed* (New York: D. Appleton:, 1873, 415.

10. Matt Ridley, *The Red Queen: Sex and the Evolution of Human Nature* (New York: Perennial Press, 1993), 118.

11. Steven Philip Kramer, *The Other Population Crisis: What Governments Can Do about Falling Birth Rates* (Washington, DC: Woodrow Wilson Center Press/Johns Hopkins University Press, 2014).

12. Razmig Keucheyan, *The Left Hemisphere: Mapping Critical Theory Today*, Gregory Elliott, trans. (New York: Verso Books, 2013), 14.

13. Curtis White, *The Science Delusion: Asking Big Questions in a Culture of Easy Answers* (Brooklyn, NY: Melville House, 2013).

14. Leo Panitch and Sam Gindin, *The Making of Global Capitalism: The Political Economy of American Enterprise* (London: Verso Books, 2012).

15. Thomas Piketty, *Capital in the Twenty-First Century* (Cambridge, MA: Harvard University Press, 2013).

Suggestions for Further Reading

For those who'd like to explore further some of the issues this book addresses, many good sources are available. I've grouped recommendations here under general topics rather than by chapter, since many of the issues come up in multiple places throughout the book. For ease of reference, the topics are listed in alphabetical order. The general topics are eugenics, genetic influences on behavior, IQ, "race," sociobiology, and the human genome. A further residual category addresses a few additional, relatively secondary issues.

EUGENICS

For eugenics, two of the best sources are Edwin Black's *War against the Weak: Eugenics and America's Campaign to Create a Master Race* (New York: Basic Books, 2003) and Daniel J. Kevles's *In the Name of Eugenics: Genetics and the Uses of Human Heredity* (New York: Alfred A. Knopf, 1985). References in these sources also offer leads for further research and inquiry into the myriad issues touching on the American eugenics movement, from Francis Galton's "positive eugenics" to their influence on Nazi policies.

GENETIC INFLUENCES ON BEHAVIOR

As to the influence of genes on behavior, a number of excellent works are available. *Not in Our Genes: Biology, Ideology, and Human Nature* (New York: Pantheon Books, 1984), by Richard C. Lewontin, Steven P. R. Rose, and Leon J. Kamin, was written before the Human Genome Project, but continues to be a very useful source. Hillary Rose and Steven Rose's *Genes, Cells, and Brains: The Promethean Promises of the New Biology* (New York: Verso, 2012) is a more recent discussion of the commercial exploitation of genetic information and misinformation, and it offers some discussion of the genome project. The book also critiques evolutionary psychology. Another source addressing the assertions of genetic determinism by evolutionary psychologists, Robert C. Richardson's *Evolutionary Psychology as Maladapted Psychology* (Cambridge, MA: MIT Press, 2007), provides an excellent critique from a philosophy of science perspective. Susan McKinnon and Sydel Silverman's edited volume *Complexities: Beyond Nature and Nurture* (Chicago:

University of Chicago Press, 2005) offers a collection of readings by specialists in various fields assessing the validity of a range of assertions advanced by evolutionary psychologists. Finally, Susan McKinnon's *Neo-liberal Genetics: The Myth and Moral Tales of Evolutionary Psychology* (Chicago: Prickly Paradigm Press, 2005) is a brief, incisive dissection of many of the questionable claims associated with that discipline.

IQ

The long-standing controversy over IQ testing, and the concept of IQ itself, has inspired a rich literature over the years. Leon J. Kamin's *The Science and Politics of I.Q.* (New York: Wiley, 1974) is an important source. Stephen Jay Gould's *The Mismeasure of Man* (New York: W. W. Norton, 1981), written by a scientist with a particular flair for addressing a general readership, has become a standard reference on the subject. More recently, Raymond E. Fancher's *The Intelligence Men: Makers of the IQ Controversy* (New York: W. W. Norton, 1987) offers a good, comprehensive summary of the lengthy debates and challenges surrounding the issue. Nicholas Lemann's *The Big Test: The Secret History of the American Meritocracy* (New York: Farrar, Straus and Giroux, 1999) discusses the broad social and historical origins and consequences of the testing industry. Some IQ discussion, of course, has involved studies of identical twins, a field in which a leading researcher was Sir Cyril Burt. Kamin's *The Science and Politics of I.Q.* raised early questions about Burt's conclusions. L. S. Hearnshaw's *Cyril Burt, Psychologist* (Ithaca, NY: Cornell University Press, 1981) is also a useful source on that topic.

"RACE"

Much of the discussion of human genetics, of course, involves "race." Jonathan Marks, in *Human Biodiversity: Genes, Race, and History* (New York: Aldine de Gruyter, 1995), does an excellent job laying out and clarifying the state of knowledge in this area. As for the fallacy of the concept of "race" and the historic persistence of the idea, many excellent works, in addition to Marks's, are available. Most recently, *The Myth of Race: The Troubling Persistence of an Unscientific Idea* (Cambridge, MA: Harvard University Press, 2014), by Robert Wald Sussman, an anthropologist who's written a good deal on the topic, and Michael Yudell's *Race Unmasked: Biology and Race in the Twentieth Century* (New York: Columbia University Press, 2014).

Other very good sources are C. Loring Brace's *"Race" Is a Four-Letter Word: The Genesis of the Concept* (New York: Oxford University Press, 2005); Jefferson M. Fish's *The Myth of Race* (Argo-Navis, 2012); Richard J. Perry's *"Race" and Racism: The Development of Modern Racism in America* (New York: Palgrave Macmillan, 2008); Bruce Dain's *A Hideous Monster of the Mind: American Race Theory in the Early Repub-*

lic (Cambridge, MA: Harvard University Press, 2002); Audrey Smedley's *Racism in North America: Origins and Evolution of a World View* (Boulder, CO: Westview, 1999); and Alexander Alland's *Race in Mind: Race, IQ, and Other Racisms* (New York: Palgrave Macmillan, 2004).

Ashley Montagu's classic work *Man's Most Dangerous Myth: The Fallacy of Race* (Walnut Creek, CA: Altamira, 1997), originally published in 1942, has been revised and updated over the years and is still available. Finally, an article by Alan R. Templeton, "Human Races: A Genetic and Evolutionary Perspective," in Robert W. Sussman, ed., *The Biological Basis of Human Behavior* (Upper Saddle River, NJ: Prentice Hall, 1999), 180–192, was a major contribution in refuting the myth that the "major races" evolved independently.

A good source on the effects of racist thinking in policies toward Native Americans is Christine Bolt's *American Indian History and American Reform: Case Studies of the Campaign to Assimilate the American Indian* (London: Allen and Unwin, 1987). Smedley also offers excellent treatment of the experience of Irish immigrants, Native Americans, and African Americans with regard to the history of racism in the United States.

SOCIOBIOLOGY

Regarding the rise and fall of sociobiology in the 1970s and '80s, a good place to start would be Arthur L. Caplan's edited volume *The Sociobiology Debate: Readings on Ethical and Scientific Issues* (New York: Harper & Row, 1978). Since it appeared during the early years of sociobiology's influence, it offers a good sense of the tone of the times. Philip Kitcher's *Vaulting Ambition: Sociobiology and the Quest for Human Nature* (Cambridge, MA: MIT Press, 1985), an extensive, fine-tuned critique. Perhaps the most influential response to sociobiology at the time was Marshall Sahlins's *The Use and Abuse of Biology: An Anthropological Critique of Sociobiology* (Ann Arbor: University of Michigan Press, 1976).

THE HUMAN GENOME

With regard to the human genome, I recommend a look at Rob DeSalle and Michael Yudell's *Welcome to the Genome: A User's Guide to the Past, Present, and Future* (New York: Wiley, 2005). The book is intended for the nonspecialist, but it offers solid, complex information and a historical perspective on past (and recent) issues. An earlier work by L. Luca Cavalli-Sforza and Walter F. Bodmer, *The Genetics of Human Populations* (San Francisco: W. H. Freeman, 1971), was a landmark in its time and still is worth a read for anyone interested in the topic. For another historic breakthrough, an article by Harvard geneticist Richard C. Lewontin titled "The Apportionment of Human Diversity," *Evolutionary Psychology* 6 (1972): 381–398, is also an important source that demonstrated that genetic diversity between any two individuals within a given population is likely to be greater than average genetic diversity among populations themselves.

OTHER ISSUES

Readers might also be interested in pursuing a number of relatively secondary issues. On claims of sociobiologists and evolutionary psychologists that the "fittest" will choose to produce more offspring because of the competition to perpetuate more of their genes, Steven Philip Kramer's book *The Other Population Crisis: What Governments Can Do about Falling Birth Rates* (Washington, DC: Woodrow Wilson Center Press/ Johns Hopkins University Press, 2014) addresses the fact that the most developed countries in Europe have been unable to maintain their existing population levels.

Marlene Zuk, in *Paleofantasy: What Evolution Really Tells Us about Sex, Diet, and How We Live* (New York: W. W. Norton, 2013), does an excellent job in assessing the recent "paleo-foods" movement and other assumptions about life in the Paleolithic, based on what we actually know about it. And for those interested in the deep historic thread of social Darwinism though much of our history, Richard Hofstadter's *Social Darwinism in American Thought* (New York: George Brazilier, 1955), originally published in 1944, is still very much worth a read.

For those interested in a general history of ideas involved in seeing human beings as subjects of scientific study, Paul Lawrence Farber's *Finding Order in Nature: The Naturalist Tradition from Linnaeus to E. O. Wilson* (Baltimore: Johns Hopkins University Press, 2000) and *Mixing Races: From Scientific Racism to Modern Evolutionary Ideas* (Baltimore: Johns Hopkins University Press, 2011) offer cogent, very readable sources.

The interplay between biological assumptions and medicine has been the subject of a good deal of analysis. Duana Fullwiley, in *The Encultured Gene: Sickle Cell Health Politics and Biological Difference in West Africa* (Princeton, NJ: Princeton University Press, 2011), discusses ways in which doctors treating sickle cell anemia patients in Senegal mistook differing outcomes among patients as indications of different genetic strains of sickle cell gene, failing to take into account differences in local modes of treatment in various communities. Also focusing on the sickle cell trait, Keith Wailoo in *Dying in the City of Blues: Sickle Cell Anemia and the Politics of Race and Health* (Chapel Hill: University of North Carolina Press, 2001) discusses the interplay between disease and "race" in Memphis throughout the past century. In this case, treatments differed because of the perceived "racial" identities of patients. Some medical practitioners resisted the idea that people with ancestries in different parts of the world often do, indeed, share higher frequencies of particular genes—in this case, the sickle cell allele, which is selected for in malarial regions.

In *Race Decoded: The Genomic Fight for Social Justice* (Redwood City, CA: Stanford University Press, 2012), Catherine Bliss discusses the general shift in attitudes among medical practitioners since the Human Genome Project. While the tendency at that time was to dismiss "racial" differences altogether, she discusses growing inclinations to take earlier "racial" categories into account in treatment. Anne Pollock's *Medicating Race: Heart Disease and Durable Preoccupations with Difference* (Durham, NC: Duke University Press Books, 2012) takes a somewhat different but related slant. Pollock focuses on differential health issues in the United States associated with being designated a member of a particular "racial" group. She points to the detrimental effects, including heightened stress, poor diet, and toxic environmental exposure, that arise as a consequence of social categorization rather than simply as results of biological ancestry.

For the most part, the sources mentioned here are very much in agreement with the positions taken in this book. Many of these issues are, or have been, quite controversial. Should anyone seek sources that take opposing points of view, the books and articles recommended here will also provide plenty of citations and references from that direction.

Index